Unmanned Aviation:
A Brief History of Unmanned
Aerial Vehicles

D0096719

Unmanned Aviation: A Brief History of Unmanned Aerial Vehicles

Laurence R. Newcome

American Institute of Aeronautics and Astronautics, Inc.
1801 Alexander Bell Drive
Reston, Virginia 20191-4344

Publishers since 1930

American Institute of Aeronautics and Astronautics, Inc., Reston, Virginia

Newcome, Laurence R.
   Unmanned aviation: A brief history of unmanned aerial vehicles/Laurence R. Newcome. –
1st ed.
      p. cm
   Includes bibliographical references.
   ISBN 1-56347-644-4
1. Drone aircraft–History. I. Title.

   UG1242.D7N48 2004
   623.74'69'0973–dc22   2004009916

Cover design by Gayle Machey

PREFACE

"Those who fail to study history are doomed to repeat it." This famous quote by the noted historian George Santayana is strikingly appropriate to the story of unmanned aviation. Over the past 85 years, robotic aircraft have been repeatedly called by the demands of war onto the stage of history to perform and perhaps to advance a step further in technology; then fade back into obscurity at war's end only to rise, phoenix-like, when the next conflict arises. With each reincarnation, lessons learned, some trivial and some significant, are lost in the ashes. Each generation of robotic aircraft designers looks around, and seeing few if any mentors from the previous generation, naturally assumes they are the first to accomplish certain feats with their newest generation of unmanned aircraft. This phenomenon is illustrated by the successful launch of Hellfire missiles from the wings of a Predator® in 2001. Claims made in the aviation media and by the program office regarding the event heralded it as the first armed unmanned aerial vehicle (UAV). Overlooked in this assertion was the 1972 history of the Firebee, which successfully launched TV-guided Maverick missiles and Hobo smart bombs against targets in the same desert that witnessed the Predator®/Hellfire debut 29 years later. Even the Firebee's accomplishment was presaged by the TDR-1s of 1944 dropping bombs on Japanese positions in the South Pacific. Such an example may be useful for establishing bragging rights but not much else; it falls into the "trivial lessons lost" category. But what of the earliest autorecovery performed in 1976 by a Ryan Model 262? How much time and money were lost in redeveloping the Pioneer's UCAR autorecovey system 20 years later and the Shadow 200's TUAV Automated Landing System (TALS) being used today? Ignorance of the history of unmanned aviation is costing us, the casual taxpayer and the UAV developer alike, time and resources by retreading old ground. If this brief history can help prevent such "reinventions of the wheel" within unmanned aviation endeavors in the future, then it will have served its purpose.

Laurence R. Newcome

CONTENTS

Chapter 1

Introduction

Unmanned aviation had its beginnings with the models built and flown by Cayley, Stringfellow, Du Temple, and other aviation pioneers as precursors to their attempts at manned flight in the first half of the nineteenth century (see Chapter 7). These models were used as the technology testbeds for larger, man-carrying versions, and in this sense they were the forerunners of manned aviation. The child's rubber band-powered toy of today represented the leading edge of aeronautics in that era, and through such models the advantages of wing dihedral and camber were revealed and studied.

In its broadest definition (i.e., aerodynamic flight without a human on board), unmanned aviation encompasses a wide range of flying devices. The genealogy of unmanned aircraft is depicted in the following figure. Starting from its roots as an "aerial torpedo," the forerunner of today's cruise missiles, its family tree has branched out to include guided glide bombs, target drones, decoys, recreational and sport models, research aircraft, reconnaissance aircraft, combat aircraft, and even exotic astroplanes—aircraft designed to fly in the atmospheres of other worlds. Unmanned aircraft in these last four branches are widely referred to today as unmanned aerial vehicles.

UAVs Defined

The term unmanned aerial vehicle or UAV came into general use in the early 1990s to describe robotic aircraft and replaced the term remotely piloted vehicle (RPV), which was used during the Vietnam War and afterward. Joint Publication 1-02, the *Department of Defense Dictionary*, defines a UAV as

> A powered, aerial vehicle that does not carry a human operator, uses aerodynamic forces to provide vehicle lift, can fly autonomously or be piloted remotely, can be expendable or recoverable, and can carry a lethal or nonlethal payload. Ballistic or semiballistic vehicles, cruise missiles, and artillery projectiles are not considered unmanned aerial vehicles.

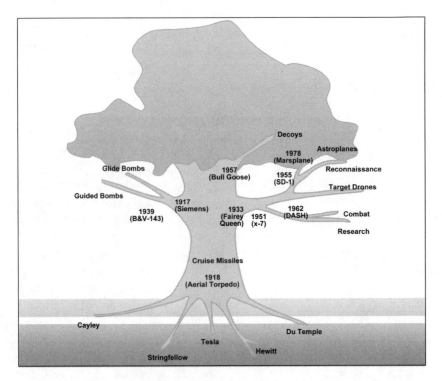

A family tree of unmanned aircraft.

In addition to distinguishing UAVs from ballistic vehicles, cruise missiles, and artillery projectiles, it also rules out gliders (which are unpowered), blimps and balloons (which float vice using aerodynamic lift), and tethered objects (which lack autonomous or remote control). This definition is being stretched by referring to artillery- and mortar-launched devices now under development, having pop-out wings and a camera in the nose, as UAVs. The terms UAV and RPV are but two of over a dozen names robotic aircraft have been known by during their existence; a chronology of these names is provided in the following figure.

The JP 1-02 definition given above is based on what a UAV is and not on what it does. Although UAVs are flown commercially in roles as diverse as precision agriculture and cinematography, their use in military circles is so overwhelmingly focused on one application—reconnaissance—that the term UAV has become virtually synonymous with reconnaissance. For this reason, the term unmanned combat air vehicle (UCAV) has come into use to distinguish UAVs designed for the strike mission from existing UAVs. The fact that this new term is not applied to existing reconnaissance UAVs that have been given a strike capability (Predator® with Hellfire missiles),

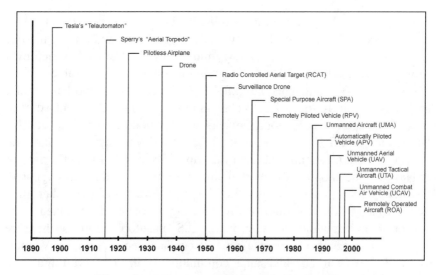

Chronology of names applied to robotic aircraft.

combined with the fact that some planned UCAVs have reconnaissance as one of their missions, argues that some other criterion is at work to win the sobriquet of UCAV. Some make the distinction that only those aircraft designed from the start for the strike mission are entitled to be called UCAVs, whereas those subsequently adapted to the strike role, such as Predator®, are not. The Loitering Electronic Warfare Killer (LEWK) UAV meets this criterion, yet it is not referred to as a UCAV, arguing that some more exclusive, unwritten qualification is being applied, perhaps having the sleek looks of a fighter aircraft. Another acronym spun off to designate a specialty UAV is MAV for microair vehicle, flying devices measuring inches in size and ounces in weight. Originally referred to as microUAVs, it was quickly pointed out the "unmanned" portion of the label was superfluous at these sizes.

CRUISE MISSILES AND TARGET DRONES

Occasionally some will refer to cruise missiles as one-way UAVs, despite the above definition specifically excluding them. To illustrate the closeness between the two, a Tomahawk flying a preprogrammed route over hundreds of miles and adjusting its speed to arrive on target at a precise time before augering in is a cruise missile, but a LEWK flying the same mission, dropping its munitions on the same targets, and returning to base, is a UAV. Just what are the distinctions among UAVs, cruise missiles, and manned aircraft? To start with the obvious, cruise missiles are expendable ordnance—one-way *weapons*, whereas unmanned aircraft are *weapon systems*. The two fundamental discriminators between the two are 1) UAVs are equipped and employed for recovery at flight's end and cruise missiles are not, and 2) while UAVs may carry munitions, the warhead carried by a cruise missile is a tailored-to-fit, integral component of its airframe.

How do UAVs and manned aircraft differ? Except for the two-letter prefix "un," they do not. If you say UAVs are controlled from the ground, so were Soviet interceptors, yet Flagons, Foxbats, and Foxhounds were manned (this also applies to US F-106s and droned QF-4s). If you say UAVs have no provisions for a pilot onboard, then the capability of QF-4 target drones or the Navy's Pelican optionally piloted aircraft to fly manned is being ignored. Was the business jet carrying golfer Payne Stewart and five others that depressurized in the first hour of flight killing all onboard a manned or an unmanned aircraft while it flew across the U.S. on autopilot? Just as there are aircraft built to carry only one person, two persons, and so forth, UAVs must eventually be seen on this same continuum as those aircraft built (or converted) to carry zero persons.

Conversions of manned aircraft to act as target drones, as alluded to above, are however considered to be UAVs. Indeed, the most enduring term for describing unmanned aircraft has been "drone", although how this appellation came to be applied to them, and specifically to target drones, is of uncertain origin. It may be traceable to British experiments in this arena in the early 1930s. Possibly it was derived from a string of word associations, beginning with the Fairey aircraft company combining its founder's name with that of a character from Milton's *Paradise Lost*, the Fairie Queene, and applying it to their unmanned variant of the Fairey IIIF scout plane in 1932. Once these few examples proved their worth as unmanned targets for the British Fleet, a large run of remotely controlled DeHavilland Tiger Moths was ordered for this purpose and, borrowing from their Fairey heritage and DeHavilland's proclivity for naming his airplanes after flying insects, renamed as Queen Bees. From Queen Bees it is a short entomological jump to the opposite sex, the drones, and the association is complete, though somewhat tortured. What is known is that Lieutenant Commander Delmer Fahrney, U.S. Navy and officer-in-charge of that service's depression-era radio-controlled unmanned aircraft project, began referring to it as the drone project in his December 1936 status report.

AIRCRAFT VERSUS AERIAL VEHICLES

Some in the UAV industry see the lack of government regulation as hobbling their efforts to expand UAVs into commercial markets by not allowing them to fly more freely in civilian airspace. As pressure has increased to regulate UAVs, the U.S. government agency responsible for doing so, the Federal Aviation Administration (FAA), introduced the term remotely operated aircraft (ROA) in 1999 because it is chartered to regulate "aircraft," and not "aerial vehicles." Aircraft require their airworthiness to be certified and their pilots to be licensed, two charter functions of the FAA. This introduced lawyers and their penchant for wording preciseness into the UAV name game. A problem arises because the FAA's definition of an ROA may not be inclusive of all types of UAVs flown by the Department of Defense (DoD). For instance, most will agree that a Global Hawk "aerial vehicle" is an aircraft and that a micro "air vehicle" is not, but just where this transition in terminology occurs is a subject of debate.

Although some may dismiss this debate as so much word fencing ("*a rose by any other name is still a rose ...*"), the distinction between what is an aircraft and what is not is at the heart of how regulators, lawyers, and, ultimately, insurers will act on UAVs, and is therefore of keen importance to the UAV industry. These definitions will determine what equipment must be carried on UAVs (and therefore their price), who is liable in mishaps involving UAVs, and how much insurers will charge UAV operators to protect them against such liability. Discussions with the FAA have helped to clarify how they categorize "device[s] ... used or intended to be used for flight in the air" (the FAA's definition of the term "aircraft") and how UAVs might relate to these categories. In general, there are three categories: aircraft (remotely operated or otherwise), which the FAA regulates; certain nonaircraft, such as ultralights, which they also regulate, though loosely; and the majority of nonaircraft, such as model airplanes, which they do not regulate. The FAA generally refers to these nonaircraft, and specifically to ultralights, as vehicles. Clearly, the military operates UAVs that are analogous to model airplanes (Dragon Eye), to ultralights (Pioneer), and to manned aircraft (Global Hawk) in characteristics and performance. How or even whether these analogies evolve into a body of regulations tailored to accommodate UAVs is an issue that concerns UAV operators and the FAA and its foreign counterparts.

TREATIES AND EXPORT POLICIES

Another arena where a precise definition of a UAV is critical is in the language used in arms control treaties and export policies. Three international agreements related to UAVs are discussed here, and it is the imprecise language of the first two that may allow them to be interpreted to apply (or not) to UAVs. These three are the Intermediate-Range and Shorter-Range Nuclear Forces (INF) Treaty, the Conventional Armed Forces in Europe (CFE) Treaty, and the Missile Technology Control Regime (MCTR).

The INF Treaty, signed in 1988, focused on removing ground-launched cruise missiles (GLCMs) and tactical ballistic missiles (Pershings) and their Soviet equivalents from European bases. It defined a cruise missile as "an unmanned, self-propelled vehicle that sustains flight through the use of aerodynamic lift over most of its flight path." Despite the concept of a recoverable UAV carrying and launching weapons being demonstrated as early as 1962, the treaty negotiators failed to conceive of a strike-capable UAV. Their oversight left open the question: are recoverable UAVs covered under the INF definition?

The 1992 CFE Treaty set limits on the number of nonnuclear weapon systems that could be stationed in Europe. It presents a similar problem, except here the definition of combat aircraft as "a fixed-wing or variable-geometry wing aircraft armed and equipped to engage targets by employing guided missiles, unguided rockets, bombs, guns, cannons, or other weapons of destruction, *as well as any model or version of such an aircraft which performs other military functions such as reconnaissance or electronic warfare*" is the issue. Are Predators® with Hellfire missiles, or even straight reconnaissance Predators®, covered under this definition?

The sale of U.S.-manufactured UAVs to foreign militaries offers the triple advantages of 1) supporting the U.S. industries and workers that make UAVs and their components, 2) potentially lowering the unit costs of UAVs sold to the services, and 3) ensuring interoperability by equipping allied forces with mutually compatible systems. Balanced against these advantages, however, is the concern that the UAV capable of carrying a given weight of reconnaissance sensors and data links on a round trip can be modified to carry an equal weight of explosives twice that distance on a one-way mission. Therefore, as the range, accuracy, and payload capacity of UAVs have overtaken those of cruise missiles and some ballistic missiles, controlling their proliferation has become a concern.

UAVs fall under the terms of the Missile Technology Control Regime, an informal and voluntary political agreement among 33 countries to control the proliferation of unmanned rocket and aerodynamic systems capable of delivering weapons of mass destruction. Table 1-1 shows which MTCR signatories are developing, manufacturing, operating, or exporting UAVs today, while Table 1-2 lists those nations who are similarly involved with UAVs but are not a party to the MTCR. Predator® and Global Hawk both fall under Category I definitions (vehicles capable of carrying 500 kg of payload to a range of 300 km) of the MTCR and were therefore subject to a strong presumption of denial for export under the original agreement. The U.S. Defense and State Departments drafted an updated interim policy to the MTCR in late 2001 to allow UAV (including UCAV) exports to selected countries on a case-by-case basis.

Table 1-1   MTCR member interest in UAVs

MTCR member nation*	UAV exporter	UAV operator	UAV manufacturer	UAV developer
Argentina	No	Yes	Yes	Yes
Australia	Yes	Yes	Yes	Yes
Austria	Yes	No	Yes	Yes
Belgium	No	Yes	Yes	Yes
Brazil	No	No	No	No
Canada	Yes	No	Yes	Yes
Czech Republic	No	Yes	Yes	Yes
Denmark	No	Yes	No	No
Finland	No	Yes	No	No
France	Yes	Yes	Yes	Yes
Germany	No	Yes	Yes	Yes
Greece	No	No	No	Yes
Hungary	No	No	No	Yes
Iceland	No	No	No	No
Ireland	No	No	No	No
Italy	Yes	Yes	Yes	Yes
Japan	Yes	Yes	Yes	Yes
Luxembourg	No	No	No	No
The Netherlands	No	Yes	No	No
New Zealand	No	No	No	No
Norway	No	No	No	Yes
Poland	No	No	No	No
Portugal	No	No	No	Yes
Russia	Yes	Yes	Yes	Yes
South Africa	Yes	Yes	Yes	Yes
South Korea	No	Yes	Yes	Yes
Spain	No	No	Yes	Yes
Sweden	No	Yes	Yes	Yes
Switzerland	Yes	Yes	Yes	Yes
Turkey	Yes	Yes	Yes	Yes
Ukraine	Yes	Yes	Yes	Yes
United Kingdom	Yes	Yes	Yes	Yes
United States	Yes	Yes	Yes	Yes

*Although not a signatory of the MTCR, Israel has pledged to abide by its guidelines.

U.S.-manufactured UAVs have been exported to numerous foreign countries over the past 40 years, from Firebees sent to Japan in the 1960s and Israel in the 1970s to Gnat 750s™ sent to Turkey in the 1990s. The Lear Astronics (now BAE Systems) R4E Skyeye®, after losing out in a U.S. Army competition, has been exported to a number of countries in Asia and Africa. Most recently, the sale of six Predators® to Italy was approved under the

Table 1-2   Non-MTCR member interest in UAVs

Non-MTCR nation	UAV exporter	UAV operator	UAV manufacturer	UAV developer
Algeria	No	Yes	No	No
Bulgaria	Yes	Yes	Yes	Yes
China	No	Yes	Yes	Yes
Croatia	No	Yes	Yes	Yes
Egypt	No	Yes	Yes	Yes
India	No	Yes	Yes	Yes
Iran	No	Yes	Yes	Yes
Iraq	No	Yes	Yes	Yes
Israel*	Yes	Yes	Yes	Yes
Malaysia	No	Yes	Yes	Yes
Morocco	No	Yes	No	No
North Korea	No	Yes	Yes	Yes
Pakistan	Yes	Yes	Yes	Yes
Romania	No	Yes	No	No
Serbia	No	Yes	Yes	Yes
Singapore	No	Yes	No	Yes
Taiwan	No	Yes	Yes	Yes
Thailand	No	Yes	No	No
Tunisia	No	Yes	Yes	Yes
UAE	No	Yes	No	No

*Although not a signatory of the MTCR, Israel has pledged to abide by its guidelines.

revised MTCR policy. Overseas interest in Global Hawk has led to one demonstration in Australia (April–June 2001), followed by a request to acquire the first 2 of as many as 20 by that government. In addition, German interest in a high-altitude endurance UAV as an option to replace its aging Breguet Atlantique maritime patrol and signals intelligence (SIGINT) aircraft by 2010 has resulted in Eurohawk, a Global Hawk optimized for the German requirements. Although miniscule in comparison to typical international sales of airliners and fighters, sales such as these mark the beginning of a sustained overseas market for American unmanned aviation.

SUMMARY

Robotic aircraft have been known by many names over the past century. UAVs are one form of robotic aircraft, and their distinguishing characteristics, besides being unmanned, are that they are powered, generate aerodynamic forces to fly, and (JP-01 not withstanding) have some means of recovery at the end of a flight. This last trait makes UAVs distinctly different from cruise missiles and other guided munitions. Having an agreed on, precise definition of a UAV is critical for use in airspace regulations, legal documents, insurance policies, arms control treaties, and export agreements. The primary international agreement impacting UAVs is the Missile Technology Control Regime, a voluntary agreement among 33 nations to limit exports of unmanned systems capable of carrying 500 kg of payload to ranges of 300 km or more. The U.S. government has interpreted this agreement so as to not impede selected exports of UAVs.

Chapter 2

BIRTH OF A CONCEPT

When a young Serbian immigrant stepped off the boat at Ellis Island in New York Harbor in 1884, he is said to have arrived with four cents in his pocket, a book of poems he had written, and his plans for a remotely controlled unmanned airplane. Nikola Tesla at age 28 was making a fresh start in life, and he had a career choice to make: pursue the life of a romantic poet or continue his rather unorthodox start in electrical engineering. Fortunately for unmanned aviation, he chose the latter.

Born in 1856 in Smiljan, a village in present-day Croatia, Tesla entered Austria's Graz Polytechnic School in 1876, despite his father's preference that he follow in his footsteps and become an Orthodox priest. There, under the encouragement of his mathematics teacher, Professor Alle, he began developing his concept for a mechanical flying machine. However, when another of his professors ridiculed his idea for transmitting electricity over long distances by alternating current, he abandoned college (and his gambling debts to classmates) and lived as a hermit in the mountainous woods near his home for some months. Emerging from his self-imposed isolation, he drifted and gambled until he learned about Thomas Edison opening a European office for his electrical products in Paris. While working there, he was rapidly educated in the leading-edge developments of the dawning new field of electrical engineering. It also introduced him to key workers in this field, American business methods, and the English language. It was this experience, along with his unmanned aircraft plans, that he brought to New York in 1884 when he met his employer. Edison was convinced the only method suitable for carrying electricity from where it was generated to where it was used was by using direct current, although this would necessitate building electrical generating plants every five blocks in New York City. Tesla proposed his alternating current approach to Edison, who, like Tesla's college professor, rebuffed him. Instead of becoming a hermit in the Adirondacks this time, Tesla turned to George Westinghouse, a businessman keen to bring the benefits and reap the profits of electrifying the city, who bought Tesla's theory of alternating current. Tesla's approach won out over that of Edison, and the modern electrical industry, indeed modern life, resulted from this partnership.

Westinghouse supported Tesla for most of the remainder of his life, although he never paid Tesla anything like the amount stated in their original contract for electrifying New York. Tesla lived in the city's finest hotels, dined nightly in its best restaurants, and was free to pursue his many other intellectual interests throughout the 1890s and into the 1900s. He never married. Inventors were the rage in turn-of-the-century American society, and the wealthy found it fashionable to patronize them and their efforts. Tesla's social circle included both inventors and wealthy investors. One regular member of this circle was Peter Cooper Hewitt, the inventor of mercury vapor lighting and another protege of Westinghouse, whom Tesla met in 1895. It was probably during these dinner conversations that Tesla described his concept for a pilotless airplane to Hewitt, who became the vector for this idea to another New York inventor, Elmer Sperry, nearly two decades later. It fell to Sperry to eventually provide the first practical demonstration of Tesla's concept.

During this same period (1890–1898), Tesla wrote prolifically for engineering journals, with some 100 articles published by or about him in *Electrical World*, 130 in *Electrical Review*, and nearly 170 in *Electrical Engineer*. The latter magazine, in which most of his work appeared, was edited by T. C. Martin, who was also a personal friend. Even so, Martin refused to publish an article by Tesla in 1898 in which he claimed he could

Nikola Tesla.

Peter Cooper Hewitt. (Courtesy of Picture History)

invent a remotely controlled aircraft that "... could change its direction in flight, explode at will, and ... never make a miss" because it was too fanciful. Tesla was well into wireless control experiments by this time, and in May of that year backed up his paper's hypothesis with a practical demonstration.

In May 1898 the Spanish–American War had just begun, and that year's Electrical Exposition drew many inventors with war-related devices to its Madison Square Garden venue. Tesla's entry was a four-foot-long boat in a tank of water that he could make stop or go, turn left or right, and blink its lights by sending out different radio frequencies. Calling it a "telautomaton," Tesla promoted it as a new form of torpedo. Its implications were lost on the military and press attendees at the exposition, who dismissed it as a trick and of no practical value. With that judgment, Tesla turned his efforts to establishing communication with the Martian civilization that Percival Lowell had supposedly just discovered, and unmanned aviation lost its best chance to beat the Wright brothers into the air by five years.

Although abated, Tesla's interest in remote control did not end with his telautomaton in 1898. A decade later, after interesting the U.S. Navy in his concept for a radio-controlled torpedo, he formed a business venture with Jack Hammond in 1912 to produce them. The torpedoes were tested at sea between 1914 and 1916, but the Navy did not pursue them further. Tesla was recognized for his accomplishments in electricity distribution with the Edison Medal before he died in 1943.

Tesla was not the only person to pursue remotely controlled weapons. Irish inventor Louis Brennan demonstrated a wire-guided torpedo, piloting it across the River Medway at Chatham, England, in 1888. Twenty years later

Tesla's 1898 "Telautomaton."

in 1908, French Artillery Officer René Lorin proposed a jet-powered flying bomb, very similar to the future German V-I buzz bomb, which would be controlled by radio from a manned escort. However, none of these approaches led to an unmanned flying device in a connected manner as did the Tesla to Hewitt to Sperry trail.

SUMMARY

In a dinner speech at the Waldorf Astoria in 1908, Tesla opined that the future of aviation belonged to dirigibles (Count Zeppelin's exploits were on newspaper front pages at the time) and that heavier-than-air aircraft would prove impractical in the long run. Such a statement seems ludicrous from the perspective of today, but it was not so unreasonable when examined in the context of his time. Dirigibles, not aircraft, bombed England eight years later, and dirigibles inaugurated passenger flights across the Atlantic in the 1920s. The commercial future of the airplane was far from certain at the time Tesla spoke, while that of dirigibles seemed bright. However wide of the mark his aviation prophesies may have been, Tesla deserves full credit for having fathered the concept of cruise missiles and the larger concept of unmanned aviation.

Chapter 3

THE CONCEPT TAKES FLIGHT

Unmanned aviation originated in the same era as manned aviation. A number of the early pioneers of flight, including Orville Wright and Glen Curtiss, contributed to the development of both. World War I was responsible for encouraging the growth of both manned and unmanned aviation, but certainly not in equal proportions. Between 1914 and 1918, manned aircraft progressed from a few hundred machines flying stunts to tens of thousands with a military purpose, whereas unmanned aircraft barely moved from the lab bench to limited production. Developments in the two fields diverged even more rapidly following World War I for a variety of reasons.

The most important reason was insufficient technology. The development of unmanned aircraft hinged on the confluence of three critical technologies, in addition to that of flight itself: 1) automatic stabilization, 2) remote control, and 3) autonomous navigation. Elmer Ambrose Sperry was the first person to attempt to address all three in a single unmanned aircraft design.

Elmer Ambrose Sperry.

Born in 1860, Elmer Sperry became a first-rank inventor in an age of great inventors and also possessed a keen business sense. His early work involved arc lamps and this may have brought him into initial contact with Peter Cooper Hewitt, a contemporary inventor of electrical lighting and fellow New Yorker. Neither Hewitt nor Sperry evidenced any interest in aviation until Sperry's work with gyroscopes for maritime applications led him to attempt to develop a gyrostabilizer for airplanes in 1909. Although Sperry's intent was to improve the safety of flight by providing a pilot with vertigo or disorientation a mechanical sense of wings level, in doing so he also solved a key technical impediment to unmanned flight: stabilized flight in the absence of a pilot's inputs. But his 30-lb gyrostabilizer, besides being excessively heavy, performed poorly when it encountered the three dimensions of flight.

With encouragement from aviation pioneer Glenn Hammond Curtiss, Sperry reattacked the problem in 1911 by using smaller gyros in the airplane's pitch, roll, and yaw axes and coupling them to the aircraft's controls by servomotors. Curtiss helped pique the interest of the U.S. Navy in this development, and they dispatched Lieutenant Gordon Ellyson to Curtiss' plant in Hammondsport, New York, in 1912 to test fly the device. Sperry continued improving the device while Curtiss tried to interest the U.S. Army in it as well. But after installing it in an Army aircraft in San Diego, Curtiss had several crashes, and Army interest quickly waned. A second series of Navy flight tests by Lieutenant Patrick Bellinger in a Curtiss seaplane was successful, but the Navy elected not to buy it, stating it was no substitute for an experienced pilot.

Sperry next mounted all three gyros on a single platform, thus providing a horizontal reference for any attitude in flight, and entered France's Airplane Safety Competition, held outside of Paris on 18 June 1914. Sperry's son, Lawrence, flew the gyrostabilized Curtiss seaplane down the Seine River, standing up with his hands raised on the first pass by the crowd, then, on a second pass, again letting go of the controls while his mechanic, Emile Cachin, climbed out on a wing. The dramatic demonstration of the utility of his father's invention won the competition's 15,000 franc prize and, later that year, the Collier Trophy for the most noteworthy achievement in aviation of 1914. The enthusiasm generated by his son's daring demonstration was short lived. War broke out two weeks later, and thoughts of gyrostabilizers gave way to armaments.

Early in 1915, Hewitt, recognizing the potential of Sperry's device as an enabler of Tesla's concept of a pilotless flying bomb, approached him with an offer of $3000 to codevelop such a weapon. Sperry readily agreed and contributed additional funds from his own company, but their joint account was soon exhausted. That same year, in response to Thomas Edison's

Glenn Hammond Curtiss.
(Courtesy United States Air Force Museum)

suggestion, the U.S. Navy formed a Naval Consulting Board to identify and pursue innovative technology for naval wartime purposes to which both Sperry and Hewitt were appointed. As chairman of the board's Mines and Torpedoes Committee and a member (along with Hewitt) of its Aeronautics

Demonstration of Sperry Gyrostabilizer, Bezons, France, June 1914.
(Courtesy United States Air Force Museum)

Committee, Sperry had little trouble gaining Navy approval and funding that October to continue developing what had come to be called an "aerial torpedo."

In September 1916 Lieutenant T. S. Wilkinson observed the first demonstration of Sperry's mechanism for guiding an aircraft over a set distance before commanding it to dive into its target from Sperry's airfield near Amityville, Long Island. For this demonstration, the system was installed in a manned aircraft, the Hewitt-Sperry Automatic Airplane, whose pilot took off before engaging the autopilot, which then leveled off at its preset altitude, flew the preplanned course, and dived at the proper distance. The pilot then recovered from the dive and returned to the airfield. Although Wilkinson reported back to his superiors that Sperry's system lacked the accuracy required to hit ships at sea, the Navy subsequently (in May 1917) recommended its further development, awarded Sperry $200,000 to proceed with it, and contributed five Curtiss N-9 seaplanes for his experiments.

The aerial torpedo program was divided into two phases, to develop initially a gyrostabilized, bomb-carrying drone with the distance gear (i.e., a fully automatic torpedo), followed by the addition of radio controls for directing the torpedo from an accompanying airplane (i.e., a controllable version). The Navy's concept of operations was to employ the torpedoes against German U-boats, their bases, and munitions factories from distances of up to 100 miles. Commander B. B. McCormick arrived at the Sperry plant in June 1917 to oversee the project. Flight tests began that September with safety pilots in the N-9s, and by November, 30-mile flights with accuracy errors of 2 miles (400 ft/mile flown) were being achieved.

Meanwhile, Sperry had engaged two subcontractors. The Curtiss Aeroplane and Motor Company was hired to develop and produce unmanned aerial torpedo airframes and the Western Electric Company was hired for their radio control system. Hewitt worked with Western Electric to develop the key component, the audion vacuum tube, which he assured Sperry would become available in time to enable radio control. As early as September 1917, however, Sperry had begun an internal effort to develop the same capability in-house, obtaining the first patent for a radio-control system (U.S. Patent No. 1,792,938) that December.

The order for six Speed Scout airframes was placed with Curtiss in October 1917, and the first was delivered just within the contract's 30-day deadline. This design was the first purpose-built unmanned aircraft. It had an empty weight of 500 lb, a range of 50 miles, a top speed of 90 mph, and was to carry a payload (bomb) of 1000 lb. The record of the 12 flights made with these six UAVs before they were all expended is shown in Table 3.1. They reflect failures due to combinations of inadequate knowledge of the aircraft's stability characteristics, gyrostabilizer failings, and launch method shortfalls.

Curtiss Sperry Aerial Torpedo.

After the first four unsuccessful flight attempts, Lawrence Sperry modified two of the aircraft for manned flight, then test flew them himself, nearly killing himself twice in an attempt to characterize their flying properties. The autopilot that had worked so consistently on the N-9 trials proved incapable

Marmon Car Launcher/Wind Tunnel.

Table 3-1   Summary of Curtiss-Sperry Aerial Torpedo flights

Flight	Aircraft date	Launch method	Result
1	1 Dec 17	Wire cable slide	Damaged wing at launch
2	1 Dec 17	Wire cable slide	Plunged into ground at launch
3	1 Jan 18	Deadweight catapult	Overturned on nose at launch
4	1 Feb 18	Deadweight catapult	Stalled, side-slipped, and crashed at launch
5	2 Feb 18	Manned (L. Sperry)	Somersaulted during taxi; wrecked
6	3 Feb 18	Manned (L. Sperry)	Autopilot rolled aircraft; recovered safely
7	6 Mar 18	Deadweight catapult	Successful flight; recovered from water
8	7 Apr 18	Deadweight catapult	Failed to rise, settled to ground, wrecked
9	17 May 18	Marmon car on rail	Lifted car wheels off track; crashed
10	5 Aug 18	Flywheel catapult	Failure; recovered
11	23 Sep 18	Flywheel catapult	Erratic 100-yard flight; crashed
12	26 Sep 18	Flywheel catapult	100-yard flight, spiraled into crash

of stabilizing the aerial torpedo after catapulting due to precession from the sudden acceleration. Finally, three means of launching the aircraft were devised, used, and discarded before a reliable catapult was developed.

To solve the problem of launching an unpiloted aircraft without upsetting its stabilizing gyroscope initially, the drone was released to slide down a wire cable. This method was succeeded by transferring the energy of a 5000-lb concrete block dropping 30 ft through a pulley and cable system to the drone. Lawrence Sperry, still not satisfied after the results of his two near-disastrous test flights, ingeniously modified a Marmon car, capable of reaching 80 mph, for use as a wind tunnel by mounting an Aerial Torpedo atop it and driving at high speed down a Long Island parkway. He next tried to use the car to actually launch the drone by fitting it with wheels from a railway hand car and mounting the combination on a nearby railroad track, only to have the airplane lift the car off the track as it approached its flying speed.

In this list of failures, however, sits a singular accomplishment: the first successful flight by a powered, unmanned aircraft, unmanned aviation's counterpart to the Wright brothers' flight 14 years earlier. On 6 March 1918, a Curtiss-Sperry Aerial Torpedo catapulted cleanly into the air, flew its planned 1000-yd flight, then dived at its preset distance into the water off Copiague, Long Island. True to the definition of a UAV, it was recovered and later reflown.

To solve the launching problem, Sperry hired Carl Norden, later of Norden bombsight fame in World War II, at the suggestion of Commander

McCormick in April 1918 to develop a flywheel-based catapult. Using a flywheel spinning at 2175 rpm, Norden's design was capable of accelerating 1950 lb to 90 kt within 150 ft. Its use on the last three flight attempts revealed structural and stability inadequacies in the Curtiss airframe. After the last prototype torpedo was expended, flight tests resumed with the N-9, and on its first launch with Norden's catapult on 17 October 1918, the now-unmanned seaplane launched without a problem, flew straight and level, and when its distance gear failed at the planned 14,500-yd mark (8 miles), continued on course out to sea at 4000 ft, never to be seen again.

With the N-9's disappearance over the horizon and the armistice at hand, the roles of Sperry and Hewitt in the development of unmanned aviation came to a close. The Navy elected to continue the development of pilotless aircraft with another company, Witteman-Lewis, and their newly designed airframe, as well as with a different autopilot manufacturer, Ford Instrument Company. Naval interest shifted from using them in the flying bomb role to that of being target drones for antiaircraft gunnery practice, and radio-control research entered a hiatus. Commander McCormick began moving the project, including Norden and his catapults, to the Naval Proving Ground at Dahlgren, Virginia, that December and by May 1919, the Sperry Company, which had surmounted two major technical challenges to unmanned flight, launching and automatic stabilization, was gone from the picture.

SUMMARY

To Elmer Sperry belongs the credit for establishing the unmanned branch of aviation. Aided by his son Lawrence and encouraged by Peter Cooper Hewitt, he built on his experience developing gyroscopes, to demonstrate the first aircraft capable of stabilizing and navigating itself without a pilot on board. Working with Carl Norden, he also devised a catapult for launching these aircraft in lieu of making a conventional takeoff. Finally, he began work on radio control, leaving its practical demonstration to his son five years later.

Chapter 4

BUT FOR ONE DISSENTING VOTE

By November 1917, Elmer Sperry was achieving consistent enough results to merit a demonstration of his gyrostabilized N-9 seaplane (with a safety pilot onboard) to a visiting group of Navy dignitaries. Major General George O. Squier, chief signal officer of the Army, attended as an invited guest of the Navy. At that time, all Army aircraft activity was overseen by the Signal Corps; it was they who had purchased the first U.S. military airplane from the Wrights in 1908. The demonstration on 21 November impressed General Squier enough to appoint a four-man board to evaluate the military potential of Sperry's aerial torpedo for Army use the following month. Within the month, three of the members recommended against pursuing the Sperry drone and, further, dismissed the idea of using planes without pilots for military purposes. The fourth member agreed that the Sperry drone was impracticable because certain of its components did not lend themselves to mass production but rejected the majority opinion that unpiloted airplanes had no potential role in war. On the strength of that single vote, the Signal Corps' Aircraft Production Board recommended the establishment of a parallel Army program for a "torpedo airplane." The lone dissenter was Charles Franklin "Boss" Kettering.

Kettering's minority report (and subsequent lobbying) was not purely altruistic. Already a well-recognized inventor and entrepreneur, he had formed the Dayton Wright Airplane Company in Dayton, Ohio, the year before from the profit earned from selling his previous enterprise, the Dayton Electrical Company, or Delco. He had promptly hired Orville Wright, who had been developing his own version of an autopilot independently of Sperry's effort, as his aeronautical consultant. Kettering, like Sperry, combined inventive genius with an aptitude for business and was already famous for inventing the cash register and the automobile self-starter. Dayton Wright had an Army wartime contract to build Liberty engines for airplanes, but none to build airplanes themselves. Kettering not only convinced the Aircraft Production Board that drones had military value and that a practical drone could be built, but that his was the company that could build it. On 8 January 1918, the Army awarded a contract to the Dayton

Wright Airplane Company for the development and production of 25 Liberty Eagle aerial torpedoes.

To his team, Kettering added the Dayton Metal Products Company for the controls; Henry Ford's chief engineer C. H. Wills and race car builder Ralph DePalma to create a new, lightweight, inexpensive engine; the Aeolian Player Piano Company for the vacuum system that was to power the controls; and Elmer Sperry for his gyrostabilization expertise. Sperry lent his expertise by commuting between Dayton and his Long Island home where his son, Lawrence, was working to get the Navy's aerial torpedo airborne.

Major C. M. Hall was appointed to oversee the development of the Liberty Eagle. Its specification called for delivering a 200-lb warhead a distance of 50 miles and to cost no more than $400 ($3600 in 2002 dollars) once in mass production. Half the size of the Curtiss-Sperry Aerial Torpedo, it was built to navigate to its target by dead reckoning using preset mechanisms, then collapse its wings and dive into its target.

The Bug, as Kettering's design came to be known, employed numerous unconventional approaches in its drive to minimize cost, ease shipping, and allow rapid production without impinging on America's already overtaxed wartime industrial capabilities. The tail surfaces were made of heavy pasteboard, the wings covered with doped muslin and paper, the engine cowling was aluminum, and the fuselage was a wooden frame with a pasteboard skin. Initially, a two-cylinder engine was used, but its vibration nearly shook the small aircraft apart. The fifth candidate engine, Wills' and DePalma's 41-hp, 151-lb, 4-cylinder design, was finally adopted. Its projected cost in production was to be $40, 10% of the planned unit flyaway price. Orville Wright's main contributions were the large dihedral of the wings (10°), which provided the lateral stability needed to prevent upsets due to side gusts, and the idea of resting the Bug on a dolly for launching so that no undercarriage would be required.

Most interesting were the Bug's innovations in flight controls, which used a combination of pneumatics, electricity, and gears. Altitude was controlled by an aneroid barometer that could be preset to a given height at which it would trip a switch to turn control over to Sperry's gyroscope to maintain its preset altitude via a link to the elevators. Direction was maintained by the gyroscope, based on its prelaunch alignment, deflecting without touching small pneumatic valves linked to the rudder. The vacuum was produced by a bellows activated by suction produced by the engine crankcase. Distance was measured by a wing-mounted anemometer tied to a pneumatically operated subtracting counter. The counter could be preset to the desired number of clicks, each corresponding to 100 yd, which on reaching zero short-circuited the engine ignition, causing the Bug to dive on its target. Short-circuiting proved to be a more elegant solution to terminating flight than the originally

Major General George O. Squier.

Charles Franklin Kettering.

planned folding-wings approach. Both the subtracting counter and pneumatic system drew on Kettering's previous experiences: the former was an earlier invention of his for cash registers, and the latter was borrowed from his home player piano, the appropriation of which reportedly invoked the wrath of Mrs. Kettering.

Because the ongoing build-up of troops in Europe put space on transport ships at a premium, the packaging of the Bug was a key design consideration. The largest piece was a wing-half, which measured 7.5 × 2.5 ft. The rear halves of the fuselages, which were pasteboard cones, could be stacked atop one another. Assembly required a screw driver and a socket wrench, involved spreading one cotter pin and tightening several bolts, and reportedly took only four and one-half minutes from opening of the box to completion of assembly.

As the flight test phase approached that summer, Kettering built a slightly larger version of the Bug, big enough to allow a pilot to fly it. The company test pilot, Howard Rinehart, began flying it that July, proving the viability of the engine then the other critical components of the Bug before it was committed to its first flight. Because the Bug was being designed for one-way flight, there was no need to throttle back for an approach and landing, hence no throttle was installed on the engine. Because of this, each of Rinehart's nearly daily test flights ended in his having to perform a dead stick landing.

Dayton-Wright Liberty Eagle.
(Courtesy United States Air Force Museum.)

General H. H. "Hap" Arnold.
(Courtesy United States Air Force Museum.)

By 1 October 1918, all was ready for the first flight except the weather, and the effect of rain at 100 mph on pasteboard would not have been a useful test point. The following day, a small amount of gasoline was loaded, the gyro powered up, the engine started, and the dolly on which the Bug rested released. As it rolled down its track gaining speed, Orville Wright was heard to comment, "She's not going fast enough." With no way to abort the launch, the Bug left the dolly when it reached the end of the track, stalled, recovered into what was described as a half-Immelmann from which it came out mere feet above the ground, stalled again, performed a second near-Immelmann this time before impacting the ground and turning end-over-end. On the evening of 3 October a quick pretest was made, using the first flight's rebuilt engine and a new airframe. This attempt flew only a hundred yards at head height before running out of fuel, but it provided confidence in the fixes made as a result of the first flight. Adding pressure was the scheduled visit the next day of Secretary of War Newton Baker, accompanied by General Squier and one of his Air Service staff, Colonel Henry Harley "Hap" Arnold, later to head the U.S. Army Air Force in World War II.

After some initial reticence, the Bug's engine sputtered to life late in the afternoon of 4 October 1918 in front of the assembled dignitaries. This time, instead of stalling immediately after launch, the engine held the Bug suspended like a helicopter as it gradually rolled onto its back, then dived toward the spectators before pulling out just above the ground and resuming a normal climb. As it climbed, its propeller torque caused it to fly in ever widening circles, which, combined with a west wind that day, caused it to drift away from the field to the east. It carried fuel for an hour's flight and soon climbed out of sight of the field, observers estimating its last seen altitude at 11,000 to 12,000 ft. Kettering ordered Rinehart to takeoff and fly after it, then land and call after he observed where it had crashed. Several carloads of people drove off in pursuit, questioning farmers as they went. Rinehart called in to report he had lost it in the gathering darkness somewhere near Yellow Springs, 25 miles from where it had taken off. Then an Army officer at nearby Wilbur Wright Field called to ask if they knew anything about a plane crashing into a barn near Yellow Springs. Kettering and Arnold quickly drove to the site of the crash and recognized the remnants as those of the errant Bug.

Now there was a new problem. Up to this point, the entire Bug project had been treated with the utmost secrecy. Kettering had moved selected employees out to a remote farmhouse to work on its design, and its construction was done inside a guarded hangar at South Field. On the days of flight tests, Kettering had scheduled them for late in the afternoon then sent the plant's employees home early. The farmer whose barn had been hit had alerted his neighbors, and they insisted on searching through the night for the missing pilot. To sustain the ploy, Colonel Arnold told the concerned farmer that the pilot had parachuted from the plane before it had crashed and had merely sprained an ankle, but the farmer was adamant he had seen the pilot fall from the plane. Only after Colonel Arnold finally offered to drive the farmer to Dayton to meet the pilot did the farmer relent. The team retrieved the scattered parts of the drone, finding all of it but one section of a wing, and assured the farmer the Army would fully reimburse him for his damages.

Despite its ad-libbed flight profile, the Bug's performance on its second outing had been impressive—nearly an hour's endurance during which it covered over 100 air miles in a stable manner and reached an estimated altitude of nearly 12,000 ft. Its performance so impressed Colonel Arnold that the following day he telegraphed Washington to recommend that the Army procure at least 10,000 and as many as 100,000 Bugs. Within five days, the Army complied with Arnold's recommendation in intent if not in quantity, ordering an additional 75 Bugs, to be delivered on or before 1 January 1919. That same week, Lieutenant Colonel Bion J. Arnold was assigned to manage the Bug's further development.

The consecutive stalls on takeoff clearly pointed to the elevator being overzealously controlled, so adjustments were accordingly made prior to the next flight. Even so, this flight bore a close resemblance to the first one, a deep stall leading into a loop, then a crash. More radical surgery was called for. The roles of the barometer and gyroscope in controlling the climb out and level off were reversed, with the barometer assuming complete control of altitude once its preset level was reached. A modest flight profile was then attempted on 22 October with over 20 Army observers on hand. The Bug climbed normally this time, leveled off at the preset 200 ft, flew the preset 1500 ft, stopped its engine, then dived on to its planned target, impacting a few feet to the right of it.

This fourth and finally successful flight was the last one made from South Field in Dayton. The plan was to now relocate the Bug's test team to Pensacola, Florida, where flights against targets up to 50 miles out in the Gulf of Mexico were to be attempted, first without a warhead to determine the drone's ballistics, then with the explosives added. But on 11 November, the armistice was signed, the follow-on production contact was cancelled, and any further testing was shelved. The blueprints, tools, and some 36 Bugs were boxed, transported to McCook Field east of Dayton, and signed over to Colonel Thurman Bane. That December, the Army totaled all the bills and concluded that $308,698.61 had been spent on developing the Bug to that point. General Motors bought the Dayton Wright Airplane Company in 1919, and Kettering was made a vice president. His Liberty Eagle gained the distinction of being the first robotic aircraft to enter (albeit abbreviated) production.

No more flights were made until 26 September 1919, when tests resumed at Carlstrom Field, Florida, where 12 Bugs had been shipped the previous month. Under the leadership of newly appointed Major Guy L. Gearhart, 14 flights were attempted, of which 10 crashed on takeoff or shortly afterward. The other four flew distances of 1.75 to 16 miles, the longest ending prematurely due to engine failure. As a result of this test series, Gearhart decided the autopilot needed a number of flights to be improved and that this testing could best be accomplished in manned airplanes. This led to a new contract in April 1920 with Lawrence Sperry and the Sperry Aircraft Company for installing these autopilots in three Standard E-1s and three Sperry Messengers for further refinement. Work on aerial torpedoes would continue, but the pasteboard Bug had made its last flight.

SUMMARY

Major General George Squier was the primary catalyst in sparking interest in unmanned aircraft within the U.S. Army after he saw the potential of these devices demonstrated in a Navy project in 1917. The Army went forward with its first UAV project on the strength of Charles Kettering's solitary vote in favor of it. Although Kettering had his prototype in the air within 10 months, the war ended in 11, short circuiting its further development. Army interest in the aerial torpedo concept continued into the 1920s, however, and work next focused on solving the challenge of remote control.

Chapter 5

TECHNICAL CHALLENGE NO. 2: REMOTE CONTROL

The difficulties encountered in simply launching a pilotless plane into the air and flying it stably shoved any plans to add the complexities of remote control of the plane to the back burner. By 1920, both the Navy's and the Army's efforts seemed to have a handle on catapulting these planes into stable flight, so attention turned to the radio engineers. Parallel projects by both the Army and Navy that were to eventually lead to remotely controlled flight were undertaken that year.

ARMY AIR SERVICE PROJECT

Lawrence Burst Sperry reentered the world of unmanned aviation after his brief absence when his firm, the Sperry Aircraft Company, was awarded an Army contract in April 1920 to build five Messenger biplanes. Designed by Alfred Verville, an aeronautical engineer employed at McCook Field (later Wright-Patterson AFB) in Dayton, Ohio, the Messenger addressed an Army Air Service requirement for a small, lightweight plane to serve as an airborne runner between the front and headquarters. As such, it had to have short takeoff and landing distances and a rugged undercarriage to allow operation from unprepared fields when necessary. The contract had two further stipulations. Sperry was to modify three of the five Messengers as well as three existing Standard E-1 scout planes for unmanned, gyrostabilized flight. If these planes validated the concept, the Army planned to produce Messenger Aerial Torpedoes (MATs) in quantity.

The Messenger weighed only 623 lb when empty and had an endurance of 4 h, a top speed of 84 kt, and a stall speed of 39 kt, making takeoff and landing easy. Sperry delivered the first Messenger five months after award, then bought it back once the Army completed its acceptance testing with the intention of marketing it in the civilian sector. Later, when the Army was slow in its contract payments, Sperry, not known for his timidity or patience, flew his Messenger from his home on Long Island to Washington, D.C., buzzed the Capitol, landed in front of it, and taxied up its steps. News cameras captured his "assault on the Hill," turning Sperry and his plane into overnight celebrities in the decade's most publicized aviation event until

Sperry's assault on the Capitol.
(National Air and Space Museum, Smithsonian Institution [SI 2002-3052].)

Lindburgh's flight five years later. The Army not only promptly paid him, but ordered 37 more Messengers, for a total of 20 unmanned and 22 manned versions, each at a price of $4000 ($38,000 in 2002 dollars).

Army flight tests began in October 1920 and, by using the Standard E-1s, focused on refining the flight control system for their eventual installation in the MAT. Over the course of these numerous flights, the automatic launching worked well, but the distance measuring system and the gyroscope that controlled azimuth were troublesome. By March the next year, the E-1 was able to consistently maintain its planned course within one degree but for only four to five minutes of flight. That May, the Army conducted a series of 10 "shots" from Mitchel Field on Long Island, 5 against a target at 8 miles near Farmingdale and 5 against another at 30 miles near Blue Point. Four of the five short-range shots ended with the distance gear initiating a dive directly over the target or on line over it. For the long-range target, two shots passed directly over the target but the distance gear failed, while the other three passed to one side of the target but the distance gear worked at the planned range.

Sperry's 1920 contract had contained $5000 performance bonuses for achieving each of the following marks:

  1) hitting a target at 30 miles 1 out of 12 flights
  2) hitting a target at 30 miles 3 times out of 12 flights
  3) hitting a target at 60 miles 1 out of 12 flights
  4) hitting a target at 60 miles 3 times out of 12 flights

5) hitting a target at 90 miles 1 out of 12 flights
6) hitting a target at 90 miles 3 times out of 12 flights and
7) making 1 successful launch and maintaining the preset heading within 2° over a distance of 1 mile
8) making 3 successful launches and maintaining the preset heading within 2° degrees over a distance of 1 mile

"Hitting the target" was defined as passing within 2° of it laterally (as measured from the launch point) and being within 2% of the total range over or under in distance. All flights claimed for the bonuses had to be flown with a government observer onboard. At the conclusion of the Long Island shots, the Army determined that Sperry had met the requirements for the bonuses under 7 and 8, but not for the other criteria. Nevertheless, it awarded him a second contract the following month (June 1921) for six additional Messengers, three of them to be configured as MATs.

In July 1921, Lieutenant Colonel Gearhart, who had been in charge of Army aerial torpedo testing since the summer of 1919, recommended that a new gyroscope, both larger and manufactured to tighter tolerances, be built to resolve the recurring problem in maintaining an accurate heading. Flight tests resumed that October, and, from the first, showed promising results. The third flight with the new gyroscope flew 75 miles from Mitchel Field to a point over New Jersey, maintaining a relatively straight line during the 119-min flight. By December, it was realized that the drones were navigating as accurately as the state of the current technology allowed and that the hurdle to improving that accuracy was the need to compensate for unpredictable changes in wind direction and velocity during the course of the flight. At this point, the possibility of adding the capability for remote control by radio was surfaced.

In May 1922, Sperry asked the Army whether installing radio controls in his MATs would violate the phrase in his contract that stated, "articles under control shall function in flight test without manual controls." The following week, the Army contracting officer stated that adding radio control would still comply with the contract's requirements. The Radio Section of the Army Air Service's Engineering Division, headed by Lieutenant Redman, developed and installed the radio equipment in several MATs during the next six weeks. On 29 June 1922, a MAT with the usual safety pilot onboard took off from Mitchel Field and flew directly over a target located 30 miles away, having received a dozen heading corrections by radio while en route. It then repeated the feat on its flight back to Mitchel Field, passing directly over the airfield, then being commanded to make a 360° turn overhead before the pilot took over and landed the plane. A second flight against a target 60 miles away was equally successful, having received 18 corrections while en route. A third flight on 30 June 1922 repeated the 60-mile target test successfully and then continued on to a target at 90 miles, also registering a direct hit. A two-seat DH-4 kept the MAT in sight and served as the controlling escort for these tests.

A week after these tests, the Chief of the Engineering Division disavowed any knowledge of the radio equipment being installed and stated his belief that the use of such equipment would be illegal under the contract. Further, he cast doubt on the military utility of these tests, saying "It would be nothing short of ridiculous to hit targets 30, 60, and 90 miles [away] and have the [MAT] target controlled by another airplane from 1 to 1.5 mile distant . . . I see no advantage in the increased range." The Air Service contracting officer apparently disagreed, because Sperry was paid $20,000 four weeks later as well as receiving personal accolades from General Billy Mitchell. A week later, four of the six MATs were shipped to Langley Field, Virginia, and the remaining two to McCook Field, Ohio, and put into storage.

In January 1923 the Engineering Division Chief defined the next challenge to fielding a militarily effective MAT as "to design, construct, and test a multiple control, i.e., one that will permit several Torpedoes to be controlled simultaneously." That March, a third contract was awarded to Sperry, this one to refurbish the six MATs for follow-on testing. Bureaucratic considerations within the Army were now raised. Was the development of radio equipment (a signaling apparatus) intended to control airplanes a responsibility of the Signal Corps or the Air Service? The question rolled about for the remainder of 1923. Finally, in March 1924, Lieutenant Koontz was allowed to purchase radio equipment and install it in the MATs.

Sperry, who had served as test pilot for much of the effort up to this point, took a break for a combined business trip/vacation to England with his family. Packed in the hold of the liner was his private Messenger, which he intended to demonstrate to Europeans. After arrival, he interested the Royal Air Force in it, even using it to drop campaign leaflets for one of the candidates in the upcoming Parliamentary elections. On the morning of 13 December 1923, Sperry took off from Rye, Sussex, headed for Amsterdam, and disappeared over the English Channel. A life-saving boat and crew found his plane, intact, floating three miles off shore but no sign of him. A month later, his body washed ashore on the English coast. The Sperry Aircraft Company closed its doors shortly afterward.

Meanwhile, service tests of the MAT conducted at Langley Field resulted in the following report card in October 1924:

> During the numerous flying tests conducted at this station, more failures resulted due to leaky pneumatic releases, broken electrical connections, precession of gyroscopes due to dirt in the bearings or parts too closely adjusted, but on the whole tests were very satisfactory and pleasing

results obtained, enough to show that the Torpedo at present is far ahead of the elementary experimental stage ... In view of the numerous possibilities of the Aerial Torpedo, it is believed that development work along this line will place the Torpedo as a foremost machine of War.

The four MATs at Langley were shipped to McCook Field in November to participate in the upcoming radio-control tests.

Captain F. H. Pritchard, who now headed the radio control project at McCook, requisitioned a JN6H airplane the previous month for his tests because the additional weight of the radio equipment was too great for the MAT to carry with its safety pilot. Test flying with the MAT resumed in December and continued until April 1925, but it was necessary for the safety pilot to assume control shortly after takeoff on every flight. By the summer of 1925, the JN6H had been modified for gyrostabilized flight as a first step, but its rudder was found to be too small, delaying installation of the radio control

Lawrence Burst Sperry.
(National Air and Space Museum, Smithsonian Institution [SI 75-16334].)

Sperry M-1/MAT Messenger.

equipment. By March 1926, Army interest and funding had faded, and the radio-controlled drone project was effectively terminated. The Messenger had a short but respectable service career. MATs were converted to manned Messengers and vice versa over the next few years, but all were eventually destroyed, either in accidents or in tests, except for one, which is now displayed in the Air Force Museum in Dayton.

NAVAL RESEARCH LABORATORY PROJECT

Leading to the second effort in developing remote control, the Navy took delivery of the first of five Witteman-Lewis pilotless aircraft in February 1919, but they proved to be just as unairworthy as the earlier Curtiss-Sperry design. Unmanned flight testing did not begin until August 1920, and then only three flights were attempted, two of them failures. The tests were supervised by Carl Norden, who had designed the flywheel catapult for the earlier drone project. He had replaced the Sperry autopilot with one of his own design in these aircraft. After the third flight in April 1921, the project became dormant due to declining Navy interest.

The Chief of Naval Operations convened a board in December 1920 to recommend whether the Navy should proceed with attempting to develop remotely control unmanned aircraft. The board recommended doing so, and

Carl Lukas Norden.

the Secretary of the Navy approved the project that January. From that quick start, activities to implement the recommendation went into slow motion. It was October before a technical team visited Dahlgren to discuss lessons learned from the inactive aerial torpedo project and to devise test procedures and delegate responsibilities. The development of the radio control system itself was made the responsibility of C. B. Mirick of the Naval Aircraft Radio Laboratory, later to be subsumed within the Naval Research Laboratory (NRL), where work did not begin until January 1922. In May, Dahlgren was directed to prepare one of the Curtiss N-9 veterans of the 1917–1918 Sperry project by July for installation of the Laboratory's radio equipment. This did not occur for another year, in July 1923. Thirty-three test flights, with Lt. John Ballentine flying as safety pilot onboard the N-9 in all cases, were made by November, when a 45-min flight involving 16 commands sent and executed during 25 min under radio control was made before senior officers of the Navy's Bureau of Ordnance. Despite the successful demonstration, a pilotless flight attempt was postponed until the following year.

Flight tests with improved radio equipment resumed in July 1924, still with Lt. Ballentine onboard. After back-to-back successful flights on 15 September, the N-9 was prepared for its first pilotless flight. The engine was started, the autopilot's gyros were spun up and unlocked, the throttle commanded up, and the seaplane released for takeoff. It flew for 40 min and executed 49 of its 50 commands before returning for a perfect landing. Coincidentally, the British had performed this same feat, unmanned flight under remote control, just 12 days earlier on 3 September 1924. Hoping to replicate this achievement in one of the two, newer Vought seaplanes allocated for these tests, the radio and autopilot equipment was transferred from the N-9 to the new plane. Check flights with a safety pilot onboard were conducted for the rest of the year.

Flights with the Vought plane resumed in July 1925, with some 30 flights being made before deciding to reattempt a pilotless flight on 11 December. Right at liftoff, the Vought porpoised several times in the manner of a pilot-induced oscillation, tearing a pontoon loose on the last bounce, which then struck the propeller before the plane nosed up and sank. The failure was attributed to the jerky inputs resulting from the push-button device used to transmit the radio commands. NRL was already developing a joystick approach, providing a much more natural piloting input for the controller, but the slow pace and erratic progress of the effort had drained Navy enthusiasm for it and no further flights were made. NRL continued improving the radio equipment over the next three years, but funding ceased and the project, and American progress in remote control, then went dormant for nearly a decade.

SUMMARY

Coincidence or not, both the Army's and the Navy's interest in developing radio-controlled flight waned with the passage of the Air Mail Act of 1925. Since 1918, Army Signal Corps pilots (later flying for the U.S. Post Office) had flown the mail, and, in line with the postman's creed, had encountered rain and snow, hail and sleet, and darkness of night. The route between New York and Chicago came to be referred to as the "graveyard run" before equipment and procedures to more safely fly in such conditions were gradually developed. When it was not dangerous, it was monotonous, droning for cold noisy hours in a straight and level line across the country and then back. It lacked the verve of "real" flying. Except for the takeoff and the landing, it could be, and eventually was, flown on autopilot. If there was a way for the pilot to stay on the ground and make the plane takeoff and land . . . . But the Air Mail Act transitioned the job to commercial contract pilots, who by 1927 had completely replaced the government in performing this service. This commercialization of a previous government function would provide the financial seed for the American airline industry, but would kill further military interest in remote-controlled flight at that time.

Chapter 6

UNITED KINGDOM

The beginnings of aviation, manned and unmanned, in England are somewhat shrouded in legend and hyperbole, largely because they are inextricably tied to an inveterate American showman, Colonel Samuel Franklin Cody. To start with, Cody was the self-adopted name of one Franklin Cowdery, born in Davenport, Iowa, in 1867. Cody never served in the military, much less achieved the rank of colonel. A cowboy in a touring wild west show who performed shooting and horse riding stunts, Cowdrey renamed himself on his 1889 marriage certificate to create an association in his audiences' mind with "Buffalo Bill" Cody (1846–1917), a famous figure from the Wild West shows of the late 1800s. His less-than-successful imitation led to his working on a horse ranch. When the ranch's owner sold some horses to an Englishman, Cody was assigned to accompany them and ensure their safe delivery.

Arriving in England in 1901, he attempted to restart his Wild West career, but by that time such shows were passé. Cody then turned to another of his interests, kite building, patenting a two-cell box kite with a set of wings added to the top of each box. England was involved in the Boer War in South Africa at the time, and Cody attempted to interest the Ministry of Defense in his kites for lifting military observers into the air. Always the showman, Cody exhibited his kites at the Alexandra Palace in 1903 and used a kite to tow himself in a canoe across the English Channel that same year. This led to demonstrations with the Admiralty later that year and with the Royal Army in 1906. After successfully flying his manned kites as high as 3400 ft, the British Army hired him as their chief instructor of kiting and put him in charge of establishing a kite unit within the Royal Engineers, which eventually became the No. 1 Squadron of the Royal Air Force. In 1905 Cody began experimenting with gliders, eventually helping to design the first military airship, Nulli Secundus, in 1907. Aboard this ship Cody helped to set a world flight record of 3 h 25 min that October.

Cody then turned to powered flight, adding a 12 hp engine to his kite and successfully flying it at Laffan's Plain in late 1907. Some claim that on one of these powered-kite flights the kite broke free, and that occurrence is credited as Britain's first unmanned aircraft flight. Cody then graduated to airplanes, building the British Army Aeroplane No. 1, and with it making the first flight of an indigenously developed British airplane on 16 October 1908. It ended in a crash, but Cody emerged uninjured. Cody repaired his airplane and went on to set a number of aviation records and win air races over the next several years, notably a prize of £5000 in the Military Trials of 1912. While giving a ride to a famous cricketer in August 1913, the airplane they were in broke apart and both were killed. In his decade of aviation pursuits, he achieved the recognition and adulation that had eluded him as a cowboy showman.

Cody was in many ways an interloper on the stage of British aviation history, a history that extended a full century before he set foot on British soil. The nineteenth century saw the foundations of modern aeronautics developed in Britain, beginning with Sir George Cayley, who is widely considered to be the first aeronautical engineer. The general public's paradigm for manned flight in 1800 was based on the man-becomes-bird myth of Daedalus and the similar drawings of DaVinci. Cayley discarded this paradigm and created a new one, one resembling the modern-day configuration of the airplane. Cayley's concept, which he engraved on a silver medallion in 1799, consisted of a large fixed wing having a camber to it and attached to a fuselage with horizontal and vertical stabilizers on

Samuel Franklin Cody.

Cody's Powered Kite, 1907.

its tail. In 1810 he published a series of three papers entitled "On Aerial Navigation," that discussed lift and drag forces, control and stability, and propulsion from a technical perspective. He also revealed in this series that he had built and flown a man-size glider at Brompton, England, in 1809—the first unmanned (albeit unpowered) airplane. After a lapse of 40 years, Cayley resumed his aeronautical experiments, building a man-carrying triplane glider in 1849. In 1853, it or a device of similar design carried Cayley's coachman on a gliding flight of several hundred yards at Brompton, after which the first-time pilot summarized his historical experience by stating, "I was hired to drive, and not to fly."

William Samuel Henson, a British lacemaker, combined Cayley's concept of the airplane with Stephenson's recent application of the steam engine to transportation into a massive advertising campaign to raise capital to create the world's first airline. His 1843 posters depicting his aerial steam carriages flying over the Egyptian pyramids and other exotic locales captured the public's imagination and sparked interest in the notion of using aviation for leisure travel. He lacked one critical item, however: an airliner. Henson's design embodied all the key elements of Cayley's earlier work and added the concepts of an enclosed fuselage and tricycle landing gear, but it only existed on paper.

A coworker of Henson's, John Stringfellow, built Henson's design, complete with its steam engine, and attempted unsuccessfully to fly versions of it between 1843 and 1847. Henson subsequently received the wrath of his investors and the ridicule of the general public, but Stringfellow's experiments in flight continued for the next 20 years. This interest contributed to the founding in 1866 of the Aeronautical Society of Great Britain, the forerunner of the Royal Aeronautical Society. Of note to unmanned aviation, Stringfellow reportedly succeeded in flying a small steam-powered model of Henson's design in 1848 and later built a model

William Samuel Henson's Aerial Steam Carriage, 1843.

triplane version that was displayed at the world's first aviation exhibition in London in 1868, thus qualifying him as the first UAV exhibitor.

Cody died on the eve of World War I, and the radical change in perspective toward aviation that that war imparted to the military is best illustrated by events in the British Army. When airplanes were first introduced as scouts in their 1911 maneuvers, the traditional Army reconnaissance force in battle, the British cavalry dismissed them as "a nuisance and frightening to the horses." Yet by 22 August 1914, only 18 days after the outbreak of hostilities, British airmen flew reconnaissance missions over Belgium in support of the British Expeditionary Force, launched the first air strike from an aircraft carrier in October 1917, and established the world's first separate air corps, the Royal Air Force, by June 1918.

Britain produced some 55,100 airplanes during the war, with production of the most expensive component, engines, being the limiting factor. The drive to develop cheaper, more producible engines led Granville Bradshaw to invent an expendable aircraft powerplant, the 35-hp ABC engine which kindled British interest in unmanned aviation. The Sopwith and DeHavilland companies, along with the Royal Aircraft Factory at Farnborough, each built prototypes of "pilotless aircraft" in 1917 that used this engine. Only the Farnborough design reached the flight test stage, but all attempts crashed during tests at Northolt, north of London. Designed by Howard P. Folland, it was equipped with radio controls developed by Archibald Montgomery Low.

After the war, the Ministry of Defence proposed three unmanned aircraft concepts. The first was a gyrostabilized, ship-launched aerial torpedo to

John Stringfellow's UAV on exhibit, 1868.

carry a 200-lb warhead, a forerunner of the Harpoon cruise missile. The second was a gyrostabilized, ship-launched aerial target for gunnery practice. The third was a radio-controlled, air-launched cruise missile with a range of 10 miles, a forerunner of the Air Launched Cruise Missile (ALCM), but this concept was never pursued.

The Royal Aircraft Establishment began experiments with the first concept in 1922 using its purpose-built RAE 1921 Target aircraft, initially in a series of launches from rails (no catapult) on an aircraft carrier, then from a high-speed destroyer. All attempts resulted in the drone plunging into the sea upon launch. Interestingly, to solve this low speed controllability problem, a radio-control system, itself a highly developmental capability, was added to provide inputs from a "pilot in the loop." This modification reversed the string of failures, and in 1924 the RAE 1921 Target went on to achieve a 39-min flight at speeds up to 100 mph, covering an equivalent range of nearly 65 miles, a significant first. Its success led to the RAE producing 12 cruise missiles of a more capable design, the Long Range Gun with Lynx Engine, or Larynx. Once its flying performance had been adequately demonstrated in flights up to 300 miles along the coast of England, five of them were sent to an RAF station near Basrah in present-day Iraq to be tested with live warheads installed. After shortcomings with the first four, the fifth Larynx flew into oblivion over the Arabian desert of present-day Iraq, the first of the numerous cruise missiles to traverse this terrain in the decades to come.

British interest in cruise missiles and in unmanned aviation in general subsided for a time following the Larynx experiments. Meanwhile, however, across the Atlantic, a man, a concept, and an event were converging to produce a domino-like chain of events that turned British unmanned aviation toward a whole new direction. The man was Billy Mitchell, a brigadier general in the U.S. Army air service, his concept was the superiority of bombers over battleships, and the event was his demonstration of this concept off the Virginia Capes in 1921 (and again in 1923). Interestingly enough, General Mitchell was focused on promoting manned aviation and nowhere in his writings showed any recognition of the potential of its unmanned cousin; it was purely an unintended beneficiary of his crusade. The first domino in this sequence was Mitchell's bombers finding, bombing, and sinking the repatriated World War I German battleship *Ostfriesland*, which neither maneuvered nor returned fire during the July 1921 attack. Mitchell repeated the feat in September 1923 by sinking the obsolete USS *New Jersey* and USS *Virginia* off Cape Hatteras, North Carolina, before his court martial in 1925. The chain he started with his 1921–1923 demonstrations led to the development of target drones by the British Fleet a decade later, which in turn led to a similar U.S. Navy effort in the mid-1930s. The Navy concept of using unmanned aircraft as targets to sharpen gunnery skills spread to the U.S. Army in the late 1930s, which built and used nearly 15,000 such drones during World War II, one model of which evolved into the first reconnaissance UAV in the 1950s. By this path, Mitchell's demonstrations ultimately led to reconnaissance UAVs.

An identical debate to that raging between General Mitchell and the U.S. Navy had also developed between the Royal Air Force, who claimed its aircraft could make easy targets of ships, and the Royal Navy, who retorted that it was the aircraft who would be at the mercy of concentrated naval gunfire. The British, however, approached the debate from the reverse direction used by Mitchell: how vulnerable was an attacking aircraft to the coordinated fire of maneuvering naval ships? The burden of proof was on the Royal Navy, as it had been previously on the U.S. Army Air Corps. Rather than the ships being unmanned targets, this approach required the airborne attacker to be unmanned, and that required the development of the world's first target drone.

To resolve this issue, it was decided to outfit an operational naval scout plane, the Fairey IIIF, with radio controls and fly it as a target to test the Royal Navy's assertion. Two of these three Fairey Queens, as they were now called, crashed during initial testing before the third one sailed with the Fleet in the spring of 1933. Catapulted from the HMS *Valiant*, the last remaining

DeHavilland Queen Bee target drone and control station.

Fairey Queen flew a lazy pattern under remote control for two hours while the Fleet shot furiously and harmlessly at it. Recovered back onboard, this one drone survived four more months of naval attempts to shoot it down before finally succumbing to these barrages. Its performance revealed the difficulty naval gunfire had destroying airplanes and complemented Mitchell's results in America, demonstrating that airplanes could find, hit, and sink large warships. Together these two demonstrations validated the air side of this interservice debate. The 1933 demonstration was the second domino in this grand sequence, and its fall led to the development and production of unmanned aircraft, on both sides of the Atlantic, dedicated to the target drone role.

The Fairey Queen's success led to a production contract for 420 DH 82B Queen Bee radio-controlled target drones, a derivative of the DeHavilland Tiger Moth trainer, built between 1934 and 1943. Air Chief Marshal Sir Michael Armitage concludes this chapter on British unmanned aircraft efforts by stating "... the fact that nearly all of them rendered very valuable service before being destroyed says more about the state of contemporary antiaircraft defences than it does about the resilience of the Queen Bee aircraft."

SUMMARY

England produced the world's first attempt at a powered unmanned aircraft with Stringfellow's steam-powered models in the 1840s. Folland and

Low of the Royal Aircraft Factory, in Farnborough, contemporaries of Sperry and Kettering, were attempting to control aircraft by radio ("pilot in the loop" flight) at the same time (1917) their American counterparts were attacking the control problem with gyroscopes (autonomous flight). As in America, interest in solving the problem of unmanned flight, and specifically radio-controlled flight, continued after the war, leading to the development of the Larynx early cruise missile and then to the use of unmanned aircraft as target drones for naval gunnery practice. This innovative application soon spread to America, making these two countries the only ones to train their military forces with radio-controlled drones in World War II.

Chapter 7

THE MANY ROOTS OF AVIATION

Much of the theoretical and empirical foundations of unmanned as well as manned aeronautics were laid in Europe during the nineteenth century. Translations from German of Otto Lilienthal's work in the 1880s and 1890s inspired the Wright brothers, who were also introduced to the earlier work of Henson and Stringfellow in England through the French-American Octave Chanute, who maintained a correspondence with several European aviation pioneers. It is fair to recognize that the Wrights' accomplishment of 1903 owed much to the successes and failures of their European predecessors. Wilbur Wright's tour of Europe in 1908, which demonstrated the performance of the Wright Flyer across the continent, also spurred a flurry of indigenous imitators, many of whom became the first native aviators in their countries.

In attempting to get themselves into the air, almost every early pioneer of aviation first built and flew—or tried to fly—a precursor model of their proposed man-carrying design. In this way, the development of the unmanned airplane occurred simultaneously with that of the manned airplane. The lessons learned in aerodynamic stability by conducting early unmanned flight, unpowered and powered, were applied directly to the betterment of manned flight attempts. From Cayley's 1804 model glider to Du Temple's clockwork-powered 1857 model and Langley's 1896 steam-powered model, unmanned flight forerunners often succeeded where their follow-on manned flight attempts did not. It is interesting to note that in recent decades, in a reversal of history, early flights of new UAV designs have occasionally employed a human pilot to ensure their initial success.

GERMANY

No history of aviation is complete without the German Otto Lilienthal, and no discussion of unmanned aviation is complete without reference to Germany's V-1 buzz bomb of World War II. Lilienthal, a mechanical engineer by profession, emphasized experimentation, focusing on the stability and control part of flight between takeoff and landing. For this he developed numerous hang gliders, made over 2500 flights with them, and published technical papers on his results. Reading these papers in 1894 sparked Wilbur Wright's interest in flight. Lilienthal was killed in the crash of one of his hang gliders in 1896.

German dirigibles began bombing raids against England in 1915 and, not surprisingly, these large, slow airships found themselves vulnerable to defending artillery and fighters. Despite conducting their raids at night, they needed the light of the nearly full moon for navigation, and this light made them discernable from the ground. Of the German Army's 35 Zeppelins, 16 were lost in combat. Their vulnerability led to the development of unmanned aircraft in Germany.

In an effort to enhance their survivability, the Siemens-Schuckert Werke began developing unmanned glide bombs in October 1915 to provide the dirigibles with a standoff capability. They were the first to employ wire-guidance, in which the glider trailed a thin copper wire after launch from the airship through which guidance corrections were sent. These electrical commands were then translated into mechanical motion by means of servo-motors connected to the control surfaces. Both of these innovations represented significant technological firsts for unmanned aviation, with wire-guidance reappearing in future generations of antitank missiles and servomotors becoming ubiquitous in UAVs. By the war's end in 1918, glides of up to 5 miles by 1-ton bombs dropped from airships had been achieved, although none were ever employed operationally.

Otto Lilienthal.

Fieseler Fi 103, or V-1 cruise missile.

With the work of Sperry and Kettering on aerial torpedoes kept under wraps during the interlude between the world wars, the Fieseler Fi 103, or V-1 "buzz bomb," introduced the public to cruise missiles and, in a larger sense, to robotic warfare. Its development stemmed from work begun in 1935 in developing the pulsejet, the simplest of jet engines, by Paul Schmidt, an inventor in Munich. In 1940, his work was combined with that of the Argus Motor Company, and their joint effort focused on developing a system for boosting takeoffs of manned aircraft. But when Luftwaffe General Erhard Milch, the head of Luftwaffe aircraft production, reviewed their progress in developing simple, cheaply produced pulsejets in 1941, he recommended their use on simple, low-cost unmanned aircraft. Thus, the concept for the V-1 was born. Production was turned over to the Fieseler Aircraft Company, and a number of Heinkel 111s were modified to air launch them. From its first launch against England on 12 June 1944 to its last on 3 March 1945, some 10,500 V-1s were launched from coastal ramps or from these bombers, with just over 2400 reaching their targets, predominantly in London. One in six V-1s were lost due to mechanical failure. Nearly 4000 were destroyed by British fighters, artillery, or barrage balloons, a 38% loss rate that justified the Germans' choice to use unmanned aircraft rather than risk their aircrews over Britain at that point in the war. Conversely, 2900 Allied aircrew members were lost defending against the V-1 attacks. A 1944 British assessment comparing the cost of waging the V-1 campaign to the Germans against the cost of its impact on the Allies concluded that the V-1 offered a 4:1 return on its investment, thus cementing the role of cruise missiles in future wars.

FRANCE

The earliest French UAV was built in 1857 by Felix Du Temple de la Croix, a French naval officer. It was a model of his patented, steam-powered, 55-ft wingspan airplane design, and it featured forward swept wings with dihedral and retractable landing gear. Propelled by a clockwork mechanism, it became the first successful powered model airplane. In 1874, Du Temple built the full-size version of his model, this time with its steam engine, and attempted to fly it with a sailor as the pilot. After running down a ramp to achieve takeoff speed, it bounced airborne briefly then faltered, falling short of the hurdle defined for sustained heavier-than-air flight by the achievement of the Wright brothers 29 years later ("... raising itself by its own power into the air in full flight, sailing forward without reduction of speed, and landing at a point as high as that from which it had started"). When Alberto Santos-Dumont, a Brazilian working in France, achieved that feat in October 1906 and became the first European to make a heavier-than-air flight, he was unaware of the Wrights' accomplishment three years earlier.

The first French reconnaissance UAV was the Nord Aviation (later Aerospatiale) R.20, a derivative of Nord's CT.20 target drone, developed in the late 1950s. The R.20 was essentially a camera on the nose of a jet engine that skimmed over hostile territory at 400 kt and as low as 700 ft along a preprogrammed route before returning its film via parachute. The French Army fielded 62 of them by 1976, eventually replacing them with the Canadair CL-289 Piver. The Israeli Air Force was briefly interested in the R.20 in 1969 but decided instead to rely on manned RF-4s for its reconnaissance needs.

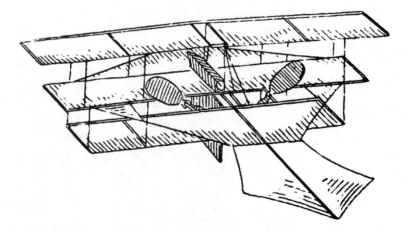

Du Temple's 1857 design.

RUSSIA

Ten years after Du Temple's near feat, Russian naval officer Alexandr Fyodorovich Mozhaisky repeated it in a steam-propelled design of his own. After three years of ground testing and with Ivan Golubev onboard, Mozhaisky's three-propeller (one tractor, two pusher) airplane rolled down a ski jump and into the air, briefly, unable to maintain or increase its altitude. Nonetheless, 50 years later, Stalinist historians credited him, as a Russian, with accomplishing the world's first powered flight, conveniently overlooking the fact that Mozhaisky had been born in what was now Finland. The flight of Boris Rossinsky at Khodynka airfield north of Moscow in 1910 is now credited with the distinction of being Russia's first heavier-than-air flight.

Immediately after World War II, the design bureau of Semyon Lavochkin developed one of the earliest mass-produced jet fighters, the La-15, known in Western circles as the Fantail. Lavochkin converted this early experience in jet-propelled flight into the La-17M target drones of the early 1950s, which were soon followed by La-17R reconnaissance drones, the first Russian UAV applied in this role. Lavochkin shifted his focus to designing intercontinental ballistic missiles (ICBMs) in 1953, leaving development of

Alexandr Fyodorovich Mozhaisky.

Russian reconnaissance UAVs to the Yakovlev and Tupelov design bureaus during the remainder of the Cold War. Reportedly, previously manned versions of the Yak-27 Mandrake, the Russian Air Force's counterpart to the Lockheed U-2, were modified to be flown unmanned in the late 1960s, making them one of the earliest high-altitude endurance UAVs.

By the early 1970s, the Russian Army's need for its own reconnaissance capability resulted in two programs, the large, high-speed Tupelov Tu-143 jet with a camera in its nose to support high-level army formations, and the smaller, slower Yakovlev Model 61 Pchela ("Bee") for regiments. Some 1000 Tu-143s were produced and deployed between 1973 and 1989, and an improved version, the Tu-243, began supplanting it in 1994. The Pchela, first flown in 1986, has been deployed to Chechnya several times since 1995, making it the first Russian UAV to be used in combat. For the future in Russian unmanned aviation, counterparts to the U.S.'s RQ-4 Global Hawk (the Sukhoi S-62), RQ-1 Predator®, and X-45 UCAV are reportedly under design.

ITALY

The Italian aviation industry had to rebuild from scratch at the end of World War II and began, logically, by undertaking the licensed production of proven foreign designs. One such effort was for U.S. Radioplane target drones, which the Italian company Meteor undertook soon after it was established in 1947. Experience gained with the Radioplane drone led to the derivative Meteor P.1 design in the 1950s, which in turn, led to the first Italian reconnaissance UAV, the Meteor P.1/R, in 1966. Although the P.1/R was flown only in line-of-sight with a television camera, the slightly larger Meteor P.2, produced until 1974, carried the same payload on beyond-line-of-sight reconnaissance missions. It was succeeded in use in the Italian Army by the familiar Mirach 10/20/26 series of twin-boom, pusher propeller designs beginning in the late 1970s, the latest evolution of which is the Mirach Falco.

JAPAN

Japan's early experiments in aviation paralleled those in the West, progressing from Ninomiya's small powered models in 1891 to the manned glider of the French attache LePrieur and Aibara in 1909, and finally to the first indigenously developed airplane, that of Nagahara in 1911. Japan attempted to field three unmanned, radio-controlled drones (cruise missiles) late in World War II. Mitsubishi's Igo-1-A was a wooden airplane carrying an 1800-pound bomb that never reached production. Kawasaki's similar, rocket-boosted Igo-1-B did reach production but not in sufficient time to see

Meteor P.2 Reconnaissance UAV.

combat; 180 were built. Tokyo Imperial University's Igo-1-C was to use an innovative acoustic sensor to allow it to home in on naval gunfire, but it too never reached production. Today, Japan leads the world in number of UAVs employed and UAV operators licensed (see Chapter 15).

Meteor Mirach 26.

Table 7-1   Selected nations' earliest known sustained flights

Country	Unmanned glider	Manned glider	Unmanned powered	Manned powered
England	Cayley, 1809	Cayley, 1849	Cody, 1907(?)	Cody, 1908
France		Ferber, 1901	Du Temple, 1857	Santos-Dumont, 1906
Germany		Lilienthal, 1891		
Japan		LePrieur/ Aibara, 1909	Ninomiya, 1891	Nagahara, 1911
Russia				Rossinsky, 1910
United States		Chanute, 1896	Langley, 1896	Wrights, 1903

SUMMARY

Europeans first developed the principles of aeronautics and, in trying to apply them to workable aircraft, flew model airplanes that could be considered the earliest UAVs. Aviation pioneers in a number of countries worldwide followed a common progression from glider to powered aircraft and from unmanned to manned flight, as shown in Table 7-1. Their technology barrier was in not having an engine with sufficient power-to-weight to enable their designs to sustain flight. American engineers, preoccupied with devising better land transportation means of conquering their continent, were latecomers to aviation, but the ingenuity of the Wright brothers and their mechanic, Charles Taylor, with his 179-lb, 12-hp engine, provided them the means to be first in flight. As in the United States, World War II spurred European development and use of unmanned target drones and reconnaissance UAVs.

Chapter 8

REGINALD DENNY AND THE RADIOPLANE COMPANY

Just as it had taken an expatriate American to introduce unmanned aviation to Great Britain, it took an expatriate Briton to reinvigorate America's interest in the same field in the 1930s. Born into a family of actors, Reginald Leigh Denny was an actor and a singer who spent his early life touring the world with his parents' stage troupe. Born in Richmond, Surrey, England, on 20 November 1891, he enlisted in the Royal Flying Corps at the age of 25, where he served as an aerial gunner during the closing year of World War I. This began his lifelong interest in, and association with, aviation. Moving to the United States in 1919, his stage credentials took him to the fledgling movie industry in Hollywood, where he appeared in numerous silent films and became one of the few actors of that era to successfully transition to "talkies." Denny also learned to fly and opened a hobby shop on Hollywood Boulevard specializing in model airplanes powered by his "Dennymite" glow plug engines. He thus became a founding father of the radio control (RC) hobby industry.

Coincident with the British turn to unmanned aircraft for use as target drones in the early 1930s, Denny incorporated Reginald Denny Industries to expand his product line of hobby airplanes into radio-controlled target drones for the U.S. military. He made two strategic hires, Kenneth Wallace Case, an electrical engineer, and Walter Righter, an engine developer, who developed the miniature radio controls and engines necessary for his drones to succeed. His first design, the Radioplane-1 (RP-1), flew in 1935 at Fort McArthur, California, but failed to produce a contract, perhaps because it crashed in front of the Army colonel sent to evaluate it. One of four commands, up, down, left, or right, each with a unique audio tone, was transmitted to the RP-1 via a rotary telephone dial. It evolved into the RP-2 in 1938, then the RP-3 by 1939, at which time the company's financial backer, the Whittier estate, withdrew its support after unsuccessful testing at March Army Airfield.

A new backer, the Collins and Powell Company, was found, and with its $75,000 investment, Denny, Whitney Collins, and Harold Powell formed the Radioplane Company late in 1939. The following year saw the emergence of the RP-4 design, whose demonstration to the Army this time led to a contract

Reginald Leigh Denny.

for what became the OQ-1 target drone. Through the World War II years, the OQ-1 evolved successively into the OQ-17 (the company's model RP-18), and the U.S. Navy also purchased these $600 drones ($6600 in 2002 dollars) under the designations TDD (Target Drone, Denny)-1, -2, and -3. In all, 15,374 copies of the RP-4 through RP-18 series were built and used to train American antiaircraft gunners during World War II.

Hollywood intervened once again in the history of Denny's drones when, as the war was drawing to a close, Denny contacted his friend and fellow actor's guild member, U.S. Army Air Force Captain Ronald Reagan, about filming Radioplane's involvement in the war effort. Reagan's unit, the 1st Motion Picture Company, was dedicated to documenting America's wartime industry, particularly Hollywood's contributions to it, and Denny was concerned the war would end without his company's contributions being captured for posterity. Reagan dispatched one of his photographers, Private David Conover, to the Radioplane Company to take pictures for *Yank* magazine on 26 June, 1945. Of the many women working on Denny's assembly line, one impressed Conover enough for him to return for additional photos and then circulate her pictures to his movie studio connections. Leaving Radioplane after his encouragement, the young woman, Norma Jean Dougherty, worked briefly as a model before graduating to film and adopting her stage name of Marilyn Monroe.

Denny's original 1935 design evolved into the OQ-19 (also known as the RP-19) by 1946, and this became the definitive model with 48,000 produced between 1946 and 1984, when it was known by its Army designation, MQM-

Radioplane OQ-1 target drone.

33. With a solid product and a long-term contract, Radioplane caught the eye of the local Northrop Aircraft Company, who acquired it in 1952. The following year, Radioplane's president, Whit Collins, became president of Northrop itself. Radioplane became the Ventura Division of Northrop in 1962 and continues today to be one of the premier producers of unmanned aircraft as part of the present day Northrop Grumman Corporation. In 1999, Northrop Grumman acquired Teledyne Ryan Aeronautical, the other major American producer of unmanned aircraft and Radioplane's chief competitor from the 1950s onward, thereby bringing together the two leaders in post-World War II American unmanned aircraft development.

Perhaps Radioplane's greatest legacy was being responsible for leading unmanned aviation into the role with which UAVs are most strongly associated today, reconnaissance in both America and in Europe. In 1955, it modified its OQ-19 Shelduck target drone into its company model RP-71 by adding film cameras. The Army introduced it into service in 1959 under the designation AN/USD-1 Observer and eventually acquired 1445 of them. An additional 32 were bought by the British Army in 1961, and it was manufactured under license in Italy. A second Radioplane design, the RP-99, progressed to the mock-up stage in 1962, but no orders were made for it by

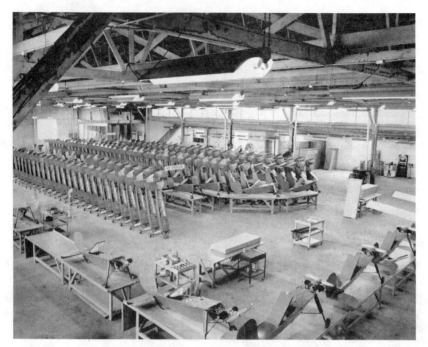

Radioplane assembly line, 1945.

Radioplane RP-71/SD-1 Observer.

U.S. services. Radioplane sold the design to Belgium's Manufacture Belge de Lampes et de Materiel Electronique SA (MBLE), who subsequently developed it as the Epervier ("Sparrowhawk") reconnaissance drone. Its first flight was made in 1965, a production order for the Belgium Army followed in 1974, and the system became operational in 1976. Subsequent attempts to sell the system, renamed Asmondee, to non-NATO countries did not materialize.

As for Radioplane's founder, Reginald Denny remained active in the Radioplane Company until its sale to Northrop in 1952. He also sold his Hollywood hobby shop at about this time and appeared in supporting roles in a few more movies (the 1966 version of *Batman* was his last) before retiring to England, where he died on 16 June 1967.

SUMMARY

Reginald Denny was the prototypical start-up for those many seedling companies in the UAV market today. Beginning with an idea, three people, and venture capital backing, he turned a lifelong interest in aviation into a hobby that is now enjoyed by thousands of radio-control airplane modelers, and then into a thriving defense business—all in the space of a decade. Most important, his basic design evolved into the world's first reconnaissance UAV, the U.S. Army's SD-1. As an actor, he had roles in over 30 movies during his career, and in turn offered the role of reconnaissance to unmanned aviation.

Chapter 9

DELMER FAHRNEY AND THE FIRST UCAV

TARGET DRONE DEVELOPMENT

A decade had elapsed since the U.S. Navy's last experiment with radio-controlled flight when Admiral William Standley, the Chief of Naval Operations, attended the London Disarmament Conference in 1935. It was while there, in conversations with his counterpart in the Royal Navy, that he learned of their use of remotely piloted aircraft (Queen Bees) as target drones and of the improvement in British antiaircraft gunnery they had produced. Upon his return to the United States, he directed the Navy's Ordnance, Aeronautics, and Engineering Bureaus to develop a similar capability for training the American fleet. Included in this direction were his nine requirements for this drone:

1) A radio-controlled seaplane
2) A speed of at least 100 mph
3) A ceiling of 10,000 ft
4) A capability for taking off conventionally or being catapulted and landing conventionally under radio control
5) A capability for straight and level flight, normal turns, climbs, glides, and entering and pulling out of dives up to 45° while under radio control
6) A capability for using the complete range of the throttle under radio control
7) A minimum control range of 10 miles from the ground control to the plane
8) Armor not required
9) Weight of onboard control equipment not to exceed that of a normal crew.

The task of converting this paper specification to reality fell to Lieutenant Commander Delmer Fahrney. Delmer Stater Fahrney grew up almost as far from the sea as a person can be and never saw it until he reported to the U.S. Naval Academy in 1916. Born on 23 October 1898 in what is now Grove, Oklahoma, but at the time was Indian Territory, Fahrney graduated from Annapolis in 1920 and later became a naval aviator. Assigned the task of developing radio-controlled aircraft in July 1936, he and his team from the

Naval Research Laboratory's (NRL) Radio Division, retracing the steps of Mirick's NRL team 12 years earlier, first designed and built the necessary radio-control equipment, then converted a number of different planes to radio control and flew them in a series of increasingly sophisticated tests over the next several years.

The first of these "NOLO" (No Live Operator) tests, conducted at the Naval Aircraft Factory in Philadelphia, involved an NRL-built, optionally piloted (i.e., manned or unmanned operation) drone called the NT, a ground control station, and a New Standard Aircraft Corporation TG-2 two-seat drone control plane. Building up from the ground station's 12-channel radio-control equipment controlling the drone's control surfaces and throttle on the ground to the TG-2 controlling the drone while both were on the ground, the TG-2 demonstrated controlling the NT on the ground from up to 25 miles away while airborne. In March 1937, both planes were flown simultaneously for the first time, with safety pilot Chief Petty Officer F. Wallace in the NT drone, while Fahrney in the TG-2 controlled the NT through a series of maneuvers. The NT was sent into a series of wild maneuvers before it was discovered that the aileron controls in the TG-2's radio unit were reversed.

Before progressing to the final step of flying the drone without a safety pilot onboard, Fahrney, feeling that landing a tail-dragger was too demanding a task for radio control at this point, wanted to convert the drone to an aircraft with tricycle landing gear. Conversion of an N2C-2 into a drone with tricycle gear was completed by October, and it was run through the complete build-up series of tests accomplished previously for the NT. The final NOLO step was attempted on 15 November 1937 after Chief Wallace had made four takeoffs and landings in it to check the radiocontrol equipment. That afternoon, Wallace assumed the role of airborne control officer in the TG-2, and Fahrney was in the ground control station. Fahrney radioed the drone's throttle to open, took off, and climbed to 200 ft before

Delmer Stater Fahrney.

Navy Research Lab N2C-2 Drone.

shifting control to Wallace in the TG-2, who put it through a series of preplanned maneuvers for the next 10 min. Fahrney then resumed control to bring it in for a landing, but a hard, nose-gear-first landing collapsed the nose wheel, and the drone skidded on its nose before coming to a stop.

A second N2C-2 and two Stearman Hammond JH-1s were assigned to the project to be modified into drones, and by late December, one of the Stearmans had been converted and fully tested. On 23 December 1937, a repeat of the 15 November pilotless test was attempted using this Stearman and completed successfully. Flight testing then paused for the winter months, during which the NRL engineers installed a feedback loop in the control station to allow the remote pilot to see the drone's actual response to commands. Flight testing resumed in April 1938, and on 27 April the original N2C-2 drone made three successful pilotless flights under both ground and airborne control. After more successes, Fahrney's team moved to San Diego and was assigned to the Fleet Utility Wing there to begin providing target drone services to the Pacific Fleet. On 24 August 1938, one of Fahrney's drones was used as a target for the gunners aboard the U.S.S. *Ranger* (CV-4), the first use of target drones by the U.S. Navy. The *Ranger*'s gunners failed to hit the drone on either of two passes, but smoke from their antiaircraft bursts obscured the drone from the controlling plane, and it crashed into the sea. It was the battleship U.S.S. *Utah*'s turn that September when the drone made a dive bombing pass at the ship instead of the level target provided to the *Ranger*'s gunners, but the *Utah*'s gunners quickly downed it anyway. The use of target drones in the Navy became standard training in 1939. The deficiencies they revealed in naval marksmanship led to improvements in optical fire control systems and encouraged the installation of radar on ships and the development of proximity fuses for antiaircraft guns, all of which proved critical to saving U.S. Navy ships and lives during World War II.

OPERATION OPTION

Although Fahrney's original orders were to develop just a radio-controlled target drone, he foresaw the potential to apply this same technology to what were now being called assault drones and recommended his project be expanded to include their development. A number of separate drone efforts began in the two years leading up to Pearl Harbor and were encouraged by the senior Navy leadership as the international situation deteriorated in 1939–1941. This support continued into the first year of the war in the Pacific, when the operational picture seemed desperate and a weapon like the assault drone offered a "Hail Mary" option in those dark days. As late as May 1942, Admiral Ernest King, the Chief of Naval Operations, directed the development, fielding, and production in quantity of an assault drone "at the earliest practicable date," based on the premise that it should be used suddenly and over a wide area to gain the element of surprise, then used continuously and heavily to forestall any countermeasures being developed against it in time. Rear Admiral Towers, Chief of the Bureau of Aeronautics, had earlier (October 1941) stipulated that the production of the drone should not impose on the already overburdened aircraft industry. One directive argued for quick production of large quantities while the other effectively hamstrung any production. Towers further suggested that instead of producing a new drone, obsolete torpedo bombers should be used, but the attack on Pearl Harbor six weeks later necessitated the use of every available plane, obsolete or not, for training and combat.

In addition to Commander Fahrney's extension of his radio-control experiments, these various projects included television and radar-based guidance systems, a homing seeker, and an "aerial ram" for air-to-air use. The last was the idea of Navy Lieutenant Robert F. Jones, commander of Utility Squadron Five (VJ-5), the target drone squadron established in March 1941 at Cape May, New Jersey, to train Atlantic Fleet gunners. His concept was to expand the role of his target drones by using them to ram enemy fighters. Admiral King approved his plan in July 1941 despite it being given negative recommendations by every reviewer en route to his desk. Despite this endorsement, Jones' Project Dog made little progress over the following year before it was rolled into the Naval Aircraft Factory's Project Fox in June 1942, which eventually produced the Gorgon missile.

By 1941, RCA had developed a 70-lb television camera and transmitter for installation in pilotless planes, such as assault drones, to assure their terminal accuracy. Flight tests of it between February and June 1941 provided a useable picture from a drone up to 30 miles distant. In tests that August, a television-guided, radio-controlled drone proved capable of hitting a target with depth charges or torpedoes on 47 out of 50 attempts, all while the drone and the target remained beyond the controller's sight. It was at this point when Admiral Towers, after being briefed on these results, produced his "not to interfere with ongoing production" edict, followed by Pearl Harbor occurring. A second series of demonstrations was held in April 1942 with equally impressive results, which led to Admiral King's "earliest practicable date" direction to establish Operation Option that May.

NRL had begun the development of a radar-guidance system for a drone in August 1941 to enable attacks in all weather or at night, two limitations of the television system. When production of 200 new-design assault drones was ordered in March 1942, they were to be capable of interchanging television and radar sensors. The first 100 drones, built by the Naval Aircraft Factory, were designated TDN-1s and the second 100, built by the Interstate Aviation and Engineering Corporation, were designated TDR-1. Both types were made of plywood and possessed low performance but could carry 2000 pounds of bombs. Building on Fahrney's lessons learned, they were equipped with tricycle landing gear. Controlled from an escorting Grumman TBF Avenger, they were guided to their targets by a TV camera in their nose linked to a 6-in. screen in the mother ship. They were to be delivered by November 1942 as the first installment of Operation Option, but the first twelve were not received by the Navy until December 1943.

Navy staff planning in June 1942 for Operation Option envisioned 18 squadrons with 162 drone control planes and 1000 drones (500 ready and 500 in reserve), as well as 10,000 naval personnel that included 1300 aviators, at a total cost of $235 million ($2.6 billion in 2002 dollars). Each of these numbers was adjusted several times during the next nine months, but the trend was consistently downward, so that by the time Admiral King approved the plan in March 1943, the operation was reduced to three squadrons of 99 control planes and 891 drones, manned by 3651 personnel. A total production run of 2000 assault drones at a peak monthly rate of 250 was planned. In the end, three Special Task Air Groups (STAG) were established

Navy/Brunswick TDN-1 Assault Drone.

Navy/Interstate TDR-1 Assault Drone.

at Clinton, Oklahoma, in the summer of 1943 and began training to fly from aircraft carriers, using the USS *Sable* on Lake Michigan.

Much of this planning had been accomplished under Captain Oscar Smith, a strong advocate for unmanned naval aviation who for security reasons had not up to this point coordinated Operation Option with the commander of the Pacific theater, Admiral Nimitz. Smith was dispatched in September 1943 to unfold this plan to Admiral Nimitz and his staff at Pearl Harbor and receive their endorsement. By this point in the war, the Navy's fast carrier task forces were successfully supporting General MacArthur's island-hopping campaign in the South Pacific and dominating the Japanese fleet in the central and north Pacific Ocean. Nimitz' conventional weapons were winning the war, so he saw no need to introduce such an unconventional weapon as an assault drone. Moreover, his carriers and airfields were fully committed, and he considered the slow speed of the drones to make them ineffective.

Smith then appealed to Admiral Halsey, commander of the South Pacific theater, whose air commander, Rear Admiral Gunther, agreed to evaluate the Special Air Task Force (SATFOR, as Operation Option was now called) drones' potential in combat in March 1944. Two squadrons of SATFOR drones sailed from San Francisco aboard the escort carrier USS *Marcus Island* (CVE 77) in May 1944 and arrived in Banika, one of the Russell Islands in the larger Solomons, just north of Guadalcanal, a few days after D-Day. Establishing themselves at Sunlight Field, the 1000 men of STAG-1 demonstrated their capability on 30 July to Admiral Gunther and Marine Major General Mitchell by diving four TDR-1s with 2000-lb bombs into a grounded Japanese cargo ship beached on Guadalcanal.

In an attempt to build on this success, Captain Smith returned to Pearl Harbor with movies of the demonstration but was again rebuffed, this time being replaced in command of SATFOR by now Commander Robert F. Jones, the first commander of the Navy's first drone squadron (VJ-5) in 1941. Authorized to conduct a 30-day combat trial of his drones, Jones deployed STAG-1 forward to Stirling Island, 50 miles south of the hotly contested Bougainville Island. They flew their first combat mission on 27 September, successfully destroying a Japanese anti-aircraft battery on a beached ship with three of four TDR-1s, each carrying a 2000-lb bomb. More missions followed against artillery emplacements, bridges, caves, tunnels, munitions dumps, and even a lighthouse. One squadron (VK-12) redeployed further to Green Island to support the assault on Rabaul, a major Japanese base. All of the drones attacked through heavy anti-aircraft fire while their Avenger mother ships orbited 6 to 8 miles away.

For the 19 October attack on Ballale Island, a new tactic was tried using a pilotless bomber. Armed with a combination of ten 500- and 100-lb bombs, a single TDR-1 dropped the bombs on gun emplacements and then headed for home. It crashed halfway back due to damage from anti-aircraft fire, but it had proven the operational validity of the unmanned combat air vehicle (UCAV). On subsequent days, TDR-1s dropped their bomb loads on targets, then reattacked by diving into Japanese cargo ships, a pilotless prelude to the kamikaze attacks that Japan began a week later. After a month of combat successes, the Navy abruptly cancelled Operation SATFOR on 27 October 1944, redeployed the Stirling and Green Island detachments back to Banika, and used the remaining TDR-1s for target practice. Of the 50 drones STAG-1 had sent into combat, 15 were lost to mechanical/technical causes, three to enemy fire, and 31 hit or damaged their targets. More importantly, not one STAG-1 aviator was lost or injured on these missions during some of the bloodiest contests of the war in the Pacific.

At the same time in Europe, another SATFOR unit, Special Air Unit 1, began Operation Anvil, the use of out-of-airframe-life bombers loaded with 24,000 lb of explosives to be flown remotely by a trailing escort aircraft into hardened German targets. The Army Air Force had abandoned its similar effort, Operation Aphrodite, using aged B-17s, due to unreliable radio control equipment. Anvil switched to the Navy's control system and outdated Consolidated PB4Y Privateers. Its drawback was that it required human pilots to take the aircraft off before turning control over to the escort once airborne; the pilots would then bail out. The first mission on 12 August 1944 exploded after the bombs were armed but before the two pilots, Lieutenants Wilford J. Wiley and Joseph P. Kennedy, bailed out. Twelve more failures spelled the end of Anvil in January 1945.

Back in Washington, Captain H. B. Temple had succeeded Captain Smith on Admiral King's staff. He quickly reviewed the assault drone project and recommended that, because of the dwindling need for these weapons, the project be restructured as a combat test effort (which it did). In March 1944 he reduced the total buy to 388 drones and the following month he recommended to Admiral King that the Navy adopt one of the assault drones under development by the Army at that time instead of developing its own. Admiral King concurred, ending the Navy's assault drone program in World War II. The SATFOR project was declassified in 1966 and recognized by the Secretary of the Navy in 1990. The last surviving TDR-1 is stored at the Naval Aviation Museum in Pensacola, Florida.

Commander Fahrney headed the Navy's efforts to develop guided missiles in the post-war years, rising to the rank of Rear Admiral before his retirement in 1953. As a result of his pathbreaking efforts in operationalizing radio-controlled aircraft, the Navy catapulted World War II-vintage, radio-controlled Grumman Hellcats with 2000-lb bombs from aircraft carriers during the Korean War to attack North Korean bridges. In his retirement, he turned—like Tesla—from unmanned aviation matters to extraterrestrial ones, becoming associated with the National Investigative Committee for Aerial Phenomena (NICAP), a group claiming that flying saucers were of extraterrestrial origin.

Summary

Modern-day UAV protagonists can easily recognize the echoes of the arguments used by today's UAV antagonists in the paperwork battles waged among the Navy staff over assault drones during the war. Given the existence of dedicated target drone units in the Navy in 1939, why did this service fail to field the same or better drones with a warhead or torpedo over the next six years? Four reasons appear.

First, the drone project management victimized themselves by requirements creep—they were beguiled by a better idea, often their own. Television guidance gave way to radar, then to radar homing, and each new innovation pushed forward the drones' entry into service date, its producibility down, and its price up.

Second, the drones were labeled as unconventional or experimental and denied access to industrial resources that could have ensured their success. Admiral Towers did not object to producing drones, just to their use of the means of production. The former would have been an unjustifiable stance; the latter was quite defensible, even patriotic, due to the ongoing ramp up to wartime production.

Third, wars tend to be "come as you are" affairs, even lengthy ones such as World War II, which lasted nearly four years. Historians have noted that all the major types of aircraft, models of tanks, and classes of ships that fought in that war had already been ordered into production before 7 December 1941 (exceptions can be argued). The target drones had relied on using cast-off airplanes for the two years leading up to Pearl Harbor, so there was no drone production base on which to build.

Fourth, the theater commander and his staff, the war fighters who inherit and employ any new weapon, were kept in the dark about its existence until they had the war on the downhill slope. Washington failed to enlist the proponency of its ultimate customer for the drones early enough. As a result, Admiral Nimitz did not perceive Operation Option/SATFOR as a long-expected guest but as a sudden and unwelcome intruder.

These four reasons reappear in subsequent conflicts and have repeatedly dogged UAV development into the present.

Chapter 10

COLD WAR, HOT MISSIONS

EMERGENCE OF THE RECONNAISSANCE ROLE

One premise of the Cold War was that the next war would be a nuclear one, and this led to the conclusion that reconnaissance missions in a post-nuclear exchange environment would be suicide for the aircrew due to the residual radiation. This assumption was validated by the experience of U.S. pilots who flew data-gathering missions over Bikini Atoll in the Pacific immediately after nuclear tests in 1946. Despite wearing lead-lined flight suits and having their aircraft washed down upon landing, radiation-related sickness occurred. Such bomb tests were also reconnoitered by de-manned B-17s under radio control. These missions introduced the "dirty" factor into considering which missions should best be delegated to unmanned aircraft.

The concept of using a robotic aircraft for reconnaissance evolved naturally during the mid-1950s from the cruise missile and target decoy roles in which they were already being used. The former demanded navigational accuracy to make them operationally viable, and the latter required a means of recovery for reuse to make them economically useful. Both attributes were necessary in a reconnaissance drone.

The reconnaissance mission itself, flown by single manned aircraft without escort or armament ("alone and unarmed") into hostile airspace, was traditionally a critical but high-attrition mission. As an illustration, during World War II, the American 3rd Reconnaissance Group lost over 25% of its pilots flying reconnaissance missions over North Africa during their initial months in theater in 1942. Compare this to a loss rate of 5.5% for American bomber crews attacking Germany in daylight in 1943–1945. Even in relative peacetime, reconnaissance flights were the consistently hot aspect of the Cold War, with 23 aircraft and 179 airmen lost between 1946 and 1990, not including another 12 aircraft and 50 airmen lost flying reconnaissance over Vietnam. Most of these "peacetime" losses went unacknowledged by both sides, although some became highly publicized, such as the Soviet shoot-down of a U-2 in 1960 and the subsequent trial of its pilot, Francis Gary Powers, causing political embarrassment for months and impacting U.S.– Soviet treaty negotiations. This danger factor, combined with the political fallout caused when airmen were captured, also helped make the reconnaissance mission a logical candidate for delegation to unmanned aircraft. Recognizing this, both the U.S. Air Force (USAF) and Army embarked on a number of "surveillance drone" programs during the 1950s (Table 10-1).

The earliest unmanned reconnaissance aircraft would have been the Northrop/Radioplane Company's B-67 Crossbow UAV had it entered operational service as planned. Originally developed by the USAF in the early 1950s as the YQ-1B high-altitude target drone, it was subsequently modified to perform the suppression of enemy air defenses (SEAD) role by the addition of a radiation-seeking sensor and warhead to home in and destroy hostile radars. A third mission, reconnaissance, was under consideration at the time the program was cancelled. Carried on wing pylons of B-47s (4) or B-50s (2), it would have been launched to penetrate high-threat areas from up to 150 miles away at 40,000 ft and 570 kt, then return to a designated recovery point, deploy a parachute, and land on airbags to allow retrieval of its camera film. However, the first flight of the SEAD version in March 1956 was followed by the program's cancellation in 1957, primarily for reasons of cost, performance limitations, and for trying to ask too many missions of a design optimized for only one; i.e., the notorious program death by "requirements creep."

The U.S. Army's SD-1 Observer could trace its design lineage directly back to Reginald Denny's original RP-1 RC model and its later versions that he eventually sold to the Army as target drones in 1939. By the end of World War II, over 15,300 examples had been produced, and it had evolved into the OQ-19 variant with the ability to enhance its radar and infrared signature to simulate a larger aircraft. It was this version (company name RP-71) to which Radioplane added cameras in 1955 and produced 1445 copies, which the Army Signal Corps deployed as tactical surveillance drones in the Army's divisions in 1959–1966. Like its target drone cousin, the SD-1 was launched using two rocket assisted take-off (RATO) bottles from a zero-length launcher; recovery was by parachute. Lacking any onboard guidance, it was flown by a ground controller who tracked it on radar, aided by a radar beacon on the vehicle, and sent it simple commands by radio. It carried either a KA-20A daylight camera that took 95 frames or a KA-39A night camera that took 10 frames during the course of its 30-min flights. In 1961, the British Army purchased 32 Observers, which served as a forerunner to that service's Phoenix UAV. The SD-1, later designated as the MQM-57, earned the distinction of becoming the first operational reconnaissance UAV.

By 1960, the U.S. Army had five surveillance drones either fielded, in flight test, or under development. The newly established Army Combat Surveillance Agency, created on 15 January 1957 in Washington, D.C., sponsored these programs, three of which were managed from Fort Monmouth,

Table 10-1 U.S. surveillance drone programs of the 1950s

UAV name	Designation(s)	Manufacturer/service	Begun	IOC	Terminated	Remarks
Crossbow	B-67 GAM-67	Northrop/Radioplane/Air Force	1954	n/a	1957	14 built
Observer (Falconer)	AN/USD-1 MQM-57	Northrop/Radioplane/Army	1955	1959	1966	1445 built
Overseer	AN/USD-2 MQM-58	Aerojet General/Army	1958	1966		35 built
Sky Spy	AN/USD-3	Republic/Army	1961	n/a	1963	
Swallow	AN/USD-4	Republic/Army	1960	n/a	1963	never flew
Osprey	AN/USD-5	Fairchild/Army	1960	n/a	1967	
Bikini	—	Republic Fairchild/Marine Corps				

New Jersey. These were the SD-2 Overseer, SD-4 Swallow, and the SD-5 Osprey (Table 10-1), none of which reached operational service.

Aerojet General's SD-2 Overseer program began in 1958 as a dedicated reconnaissance UAV capable of carrying a variety of "plug and play" sensors that could be changed out for one another. These sensors consisted of a 125-mm or a 70-mm film camera, flares for use with them for nighttime photography, an IR sensor, a side-looking airborne radar (SLAR), and the plumbing for dispensing chemical or biological agents from underwing tanks. The 125-mm film had to be developed after recovery, but the 70-mm film was developed while in flight, scanned, and its imagery transmitted to its ground sensor terminal with only a 1-min delay. The sensor data link was also capable of transmitting the IR or SLAR imagery back in real time. The command data link was a continuous transmission of spread spectrum pulses from a master and repeater ground station. The time differences of the pulses returning from the aircraft were used for its navigation scheme, called translateration, which worked much like differential GPS does today, and the control inputs were embedded in the relative security of the spread spectrum stream. Accuracies of 5 ft out to 50 miles were eventually demonstrated using this method, but not before systemic navigational shortcomings led to the program's cancellation in 1966. The SD-2 relied totally on a preprogrammed route without external references to bring it over its intended targets, so a precise navigational capability was key to the system's success or failure. Poor reliability was also a factor. The SD-2 was launched off a rail using a RATO bottle to boost it to flying speed, and this kick start was anathema to its vacuum tubes. Some 35 Overseers were built but never operationally

Army/Aerojet General SD-2 Overseer.
(Courtesy of Research Library at U.S. Space and Rocket Center, Huntsville, Alabama.)

fielded. It was a noteworthy design for the technology stretch it attempted, in guidance, navigation, and communication, but which proved to be a span too far to bridge for its time.

The SD-4 and SD-5 were competing designs for long-range, high-speed reconnaissance support for targeting tactical ballistic missiles by Army corps. Both were subsonic, delta-winged, turbojet-powered designs. The SD-4 Swallow, built by Republic, never reached the flight test stage and was cancelled in 1963 for cost. Fairchild's SD-5 Osprey had the advantage of being a spin-off from the Air Force Strategic Air Command's Bull Goose program to develop a ground-launched, long-range decoy meant to simulate B-52s on Russian radars and draw air defenses to them and away from the real bombers. It first flew in 1960 and later demonstrated 4-hr flight durations over preprogrammed routes relying solely on its onboard inertial navigation system. Accelerated to flying speed by a 40,000-lb thrust rocket booster, it recovered from flight with a combination of parachutes and airbags. It, too, was cancelled in 1963, which was its originally planned operational date, for cost reasons.

The fifth Army UAV project at this time was the Republic SD-3 Sky Spy. Although it never saw operational service, the innovative configuration of the Sky Spy, a twin-boom shielding a pusher propeller, proved to be prescient for future Israeli and American UAV designs, such as Scout, Pioneer, and Shadow. Intended for medium-range reconnaissance missions, Sky Spy was designed with interchangeable noses for film, infrared, or radar payloads.

Finally, the U.S. Marine Corps had instituted a program for a small, battalion-level drone in this same timeframe. Developed by Republic under an Office of Naval Research contract, the Bikini (General Wallace M. Greene, Jr., Commandant of the Marine Corps at the time, explained its name was due to it being "a small item that covers large areas of interest") was a

Army/Republic SD-4 Swallow.

Army/Fairchild SD-5 Osprey.

60-lb UAV with a 30-min endurance. Its two-man team drove the system around in a jeep and trailer, charging its pneumatic launcher with the same compressor used to recharge the unit's flame thrower tanks. In addition to reconnoitering enemy activities, Bikini was to be used to check the battalion's camouflage discipline. Operational evaluations began in 1963 at Twenty Nine Palms Marine Corps Base in California, where 327 flights were made by 20 drones before the program, begun in 1959, was canceled in 1967 for technical shortcomings. Interestingly enough, the concept of employment generated for the Bikini was recycled for the Close Range UAV program in 1992.

AUTONOMOUS NAVIGATION

In addition to the renowned arms race between the United States and the USSR during the Cold War, there was also a technology race internal to the United States during the first decade of that era. This was a race for dominance between two emerging classes of robotic weapons: intercontinental ballistic missiles (ICBMs) and unmanned aircraft to serve as intercontinental cruise missiles. The race had been presaged by the appearance of the German V-1 cruise missile and V-2 ballistic missile in the closing year of World War II's European campaign. It would result in the solution to unmanned aviation's third and last remaining technical hurdle, autonomous navigation.

The competition arose because of the weight of early nuclear weapons. Early ICBM designs had insufficient "throw-weight" to carry one of these

Army/Republic SD-3 Sky Spy.

devices the required distance of 5000 miles. However, because their speeds averaged some 10,000 mph, flight times would be on the order of 30 min, meaning navigation errors would have very little time to accumulate and affect the ICBMs' accuracy. On the other hand, because their aerodynamic lift supplemented their thrust, unmanned aircraft were capable of carrying the weight of early bombs, but their flight time at 500 mph meant hours of accumulating navigational error that would degrade their accuracy. The race became one between nuclear physicists reducing the weight of the bomb and guidance engineers developing a means of highly accurate navigation without human inputs. If the former arrived at a solution first, the ICBM

Marine prepare "Bikini" drone for launching   Compressed air "blast-off."

Marine Corps/Republic Bikini.
(Reprinted courtesy of the Marine Corps Gazette. Copyright retained by the Marine Corps Gazette.)

concept won; if the latter, cruise missiles would become the defense paradigm. The ICBM faction won, but UAVs were the benefactor of this cold war technology push to develop self-contained, highly accurate navigational systems.

Navigational accuracy was critical to both types of vehicles, just more so in the aircraft case. Various schemes were devised to provide this accuracy and thus improve weapon delivery accuracy over intercontinental distances. These schemes were of four general categories: active or passive and dependent or independent, with active/passive depending on whether the navigation system was required to emit and dependent/independent on whether its functioning required externally generated inputs or was entirely self-contained. Examples are shown in Table 10-2.

For military purposes, vulnerability was reduced by using passive rather than active and independent rather than dependent systems; thus inertial navigation systems (INS) were favored. The only problem was that while inertial navigation was an ideal concept in the late 1940s, it needed someone to convert the concept into useable hardware. The person who stepped up to that task was Charles Stark Draper, a professor of aeronautics and astronautics at the Massachusetts Institute of Technology (MIT).

Born in Windsor, Missouri, in 1901, "Doc" Draper earned undergraduate degrees in psychology and electrochemical engineering before earning his doctorate in physics in 1938. Sponsored by the Sperry Gyroscope Company, he developed gyroscopically stabilized gunsights for naval gunnery that were installed on U.S. Navy ships throughout the war and used to great effect in directing gunfire against attacking Japanese aircraft. This led him to evolve the theory, invent the technology, and lead the effort to develop, manufacture, and field inertial navigation systems for aircraft, submarines, and space vehicles. He developed the first INS for aircraft navigation in 1949, and the first fully inertial INS, the Space Inertial Reference Equipment (SPIRE), in 1953. Enhanced versions of SPIRE were used in the guidance systems of the Thor, Polaris, Poseidon, and Trident ICBMs, and, later, in the Apollo moon landing program. Prior to this time, the early developers of unmanned aircraft had had to rely on clockworks and timing gears to estimate distances flown, combined with gyrocompasses for heading.

Table 10-2   Categories of navigation systems

	Dependent/External	Independent/Internal
Active	LORAN (1940s)	Radar (1940s)
	Transit (1960s)	Terrain Comparison (1950s)
	Global Positioning System (1980s)	
Passive	Star Tracker (1950s)	Inertial Navigation System (1940s)

Severe turbulence or aircraft maneuvers easily upset these mechanisms and made the vehicle's navigation inaccurate. In addition there was no way to update these mechanisms once launched; only radio commands could be used to retake control of the aircraft. Draper's inertial systems solved these problems and enabled robotic flight to accurately navigate without dependence on external inputs.

An early INS was developed for the Snark cruise missile, and an astro-tracker was added when the early INS's accuracy proved insufficient, but its complexity and weight (1000 lbs) compounded rather than alleviated the problem. Together they provided an accuracy of 1 mile in 100 of travel. Although these weapons were being developed to wage war at intercontinental distances, just which continent was sometimes in question. A Snark, launched eastward from Florida on a test flight in 1956, disappeared from tracking radars, never to be seen again until 1982, when a Brazilian farmer came upon it while clearing land. A hyperbolic radio navigation system (same principle employed by LORAN and in today's Global Positioning System) was developed for a concurrent, smaller cruise missile, Matador, but

Charles Stark Draper.

Inertial Navigation System.

limited its useful range and was vulnerable to jamming. A terrain matching system was developed and added to overcome these shortcomings, but the lack of reliable radar maps of the target areas shackled its contribution.

All four of these automatic navigation concepts were essentially ahead of their time in the 1950s, but by the 1960s, astrotrackers were guiding interplanetary probes to Mars and Venus, and INS accuracy was sufficient to guide man to and from the moon. By the late 1970s, computer size had shrunk and satellite support increased to the point that terrain matching could be a reliable adjunct to the INS in cruise missiles. Satellite constellations emitting radio signals for navigation appeared in the 1960s (Transit) and matured in the 1980s (GPS).

Today, UAVs rely on combinations of INSes using ring laser gyros and GPS receivers that are updated by satellite radio signals to maintain navigational accuracies under 20 ft during missions over 24 h in duration. It is directly due to the early work of Elmer Sperry in automatic stabilization, Lawrence Sperry, C. B. Mirick and Delmer Fahrney's team in radio control, and Charles Draper in autonomous navigation that the Global Hawk UAV was able to fly unerringly from California to Australia nonstop and touch down within feet of the runway centerline upon landing in April 2001.

SUMMARY

The competition between ICBMs and cruise missiles to be the U.S.'s intercontinental weapon of choice in the opening years of the Cold War led directly to overcoming the last major hurdle for unmanned aviation and autonomous navigation. Credit for this signal accomplishment is due to Charles Stark Draper and his team at MIT. The 1950s also saw the first development and fielding of unmanned aircraft as reconnaissance UAVs, followed by an aggressive effort by the U.S. Army Signal Corps to develop a family of short-, medium-, and long-range reconnaissance UAVs in the early 1960s. Although the Army did field over 1400 small tactical surveillance drones, all of their other programs were eventually canceled as bills for the Vietnam War rapidly began to increase in the mid-1960s.

Chapter 11

Vietnam: The Coming of Age

The Vietnam War was notable in two regards with respect to unmanned aircraft. It was the first war in which reconnaissance UAVs were employed, and it was also notable for the ubiquity of the drones' use throughout the war—an average of one drone mission was flown each day of this lengthy war. Three American UAVs, the Air Force/Ryan AQM-34 Lightning Bug, the Navy/Gyrodyne QH-50 Drone Antisubmarine Helicopter (DASH), and the Air Force/Lockheed GTD-21 had roles in the southeast Asia (SEA) conflict.

AQM-34 Lightning Bug

From its first operational mission over SEA (actually over China) on 20 August 1964 until the last mission in June 1975, 23 versions of the Ryan AQM-34 Lightning Bug flew 3435 sorties in support of the Vietnam War. In addition to conducting imagery reconnaissance, these sorties included electronic and communication intelligence (ELINT and COMINT) collection, decoy, and leaflet-dispensing missions. Also, variants were developed to carry chaff and electronic countermeasures (ECM) systems for the SEAD mission under the Combat Angel program, as well as to launch a variety of air-to-ground munitions, including AGM-65 Maverick and Stubby Hobo TV-guided missiles and 250- and 500-lb general purpose bombs. This 1971 project, Have Lemon, made the Fire Bee II one of the earliest explorations of the military utility of strike UAVs and a forerunner to the Air Force and Navy UCAV programs, also having primary missions of defense suppression, 30 years later.

The 1964 Lightning Bug was a second-generation Ryan reconnaissance drone, having evolved from Ryan's earlier Q-2C Firefly reconnaissance drone, of which four were built. Both were euphemistically referred to by the Air Force as "special purpose aircraft" (SPA). The Q-2C had itself evolved from the Ryan Q-2 Firebee target drone in 1962, just months before the Cuban Missile Crisis in October. The Q-2C was quietly developed in response to the Soviet Union shooting down Francis Gary Powers and his U-2 in May 1960. On 27 October 1962, Major Rudolph Anderson and his

U-2 were shot down over Cuba, and the two remaining Fireflys were scheduled to overfly the island a few days later on what would have been the first combat mission for a reconnaissance UAV. Their mission was scrubbed at the last moment (the C-130 with the two drones on its wing pylons reportedly had its engines running), the Cuban crisis itself defused, and the reconnaissance drone capability that had been developed under a classified Air Force program was kept under security wraps for another two years.

When the Ryan AQM-34 did go to war in August 1964, it did so by flying over southeastern China, from Hainan Island to opposite Taiwan, where it was recovered. A typical mission saw the C-130 with its two SPAs, a primary and a backup, take off from Kadena Air Base on Okinawa at dawn, launch the SPA off China at noon on its two- to three-hour preprogrammed mission (no en route corrections were made), then return to Kadena. Meanwhile, the SPA returned over Taiwan, where it deployed a parachute over a predesignated location and a second C-130 retrieved it and flew it back to Kadena that evening. The mission's film was removed and packaged at the Taiwan recovery site, put onboard a courier jet, and flown to Offutt AFB in Omaha, Nebraska, for interpretation.

In October, the operation, called Operation Blue Springs, moved to Bien Hoa Air Base, 20 miles north of Saigon in South Vietnam, which served as the home of the SPAs until July 1970 when it was moved to U-Tapao, Thailand, for the remainder of the conflict. AQM-34 missions from Bien Hoa continued to cover China and were complemented by manned U-2 flights

Ryan AQM-34 Lightning Bug.
(Courtesy United States Air Force Museum.)

covering North Vietnam. This breakout was driven by the desire to avoid a repeat of the Francis Gary Powers incident with politically sensitive China and the absence of any surface-to-air missile (SAM) threat over North Vietnam. This approach was validated in April 1965 when China put the remains of three AQM-34s they had shot down on display in Peking, and it was essentially a one-day story in the Western newspapers.

When SA-2 high-altitude SAMs were introduced into North Vietnam later in 1965, the U-2s began flying stand-off missions, and the AQM-34s took over most of the now high-risk penetration flights into North Vietnamese airspace. If a SPA failed to cover the required target of interest, a manned reconnaissance aircraft, such as an RF-101 Voodoo, would typically be tasked to streak in at low altitude over the target the next day.

Being the new kid on the reconnaissance block, and an unmanned one at that, the AQM-34 had to fight for acceptance on two fronts, from other reconnaissance pilots and from those pilots assigned to fly it, both of whom initially perceived it as somehow diminishing to their jobs. Two significant events in October 1965 helped mollify this perception. First, a joint U-2/RB-47/AQM-34 mission was conducted in which the drone purposely drew SA-2 fire while the U-2 and the RB-47 stood off to record and report on the intercept tactic used. That cooperation for a common cause, plus the visual impact of watching what a SAM could do to an aircraft, helped convert the reconnaissance pilots. Second, the AQM-34 team began referring to their SPAs as "remotely *piloted* vehicles" (RPVs), which helped convince their own aircrews that they still had a pilot's function.

Nevertheless, as AQM-34s flew more reconnaissance sorties and more varied missions, they increasingly came to be seen as competitors by some pilots. This generated what was called the "white scarf mentality" and the aviators who espoused it were referred to as the "white scarf mafia." Within the SPA's parent organization, the USAF's Strategic Air Command (SAC), drones were seen as politically competitive with manned U-2s and SR-71s. Although SAC officers recognized the complementary capabilities of the drones, U-2s, and SR-71s, they were concerned someone in Congress, applying media-mathematics, would compare the cost of an SR-71 flight hour (about $250,000 in 1971) with that of the SPAs and try to cancel the former.

Exacerbating this divisive attitude during Vietnam was the long-standing rivalry within the USAF between the strategic (bomber) side of SAC and the tactical (fighter) side of Tactical Air Command (TAC), whose 7th Air Force was flying most of the combat missions over Vietnam. When the operations officer of the RF-4C tactical reconnaissance squadron based at U-Dorn, Thailand, quietly asked SAC's drone unit to cover a target that was proving too hot for his aircrews, he was told to send his request through SAC.

"SAC?" he replied, "If I ever asked SAC for help, I'd be laughed out of my squadron." The following drone mission was quietly modified to accommodate the RF-4's unruly target.

The concept of operations in Vietnam evolved from flying missions at high altitude to low altitude, from parachute recoveries—often in hostile rice paddies—to recovering the drones in midair by helicopter (the Mid-Air Retrieval System—MARS) and from flying the film 9000 miles to Omaha for exploitation to performing this function 20 miles away in Saigon. Beginning in 1966, MARS was used on 2745 recovery attempts and was successful on 2655 (nearly 97%) of them. A ground control station was also established at Da Nang to aid in control and recovery; otherwise the drone was controlled from the DC-130 mother ship. The pace of operations increased from two missions a week in late 1964 to nearly two missions a day when the North Vietnamese Army crossed into South Vietnam in 1971 to four missions a day during Operation Linebacker, the bombing of North Vietnam by B-52s in 1972.

Over the course of the entire war, from 1964 to 1975, the Lightning Bug compiled the following statistics: 1016 AQM-34s flew 3435 combat sorties during which 544 were lost, approximately two-thirds to hostile fire and one-third to mechanical malfunctions, for an overall mission success rate of over 84%. Three of the losses were attributed to radio interference from friendly outposts operating on the drone control frequency. Over 100,000 ft of film were taken and recovered.

The Lightning Bug was a versatile design, and Vietnam was a fertile proving ground for exploring new roles for UAVs. Besides being the first UAV used as a decoy in combat (March 1966), the Lightning Bug was adapted to suppress enemy air defenses (SEAD) with chaff (August 1968), collect communication intelligence (COMINT; February 1970), launch Maverick missiles against fixed targets (December 1971), and drop leaflets for psychological operations (July 1972). New operational concepts and sensor technologies were also explored. The Navy demonstrated the capability to operate them from an aircraft carrier in 1969–1970, and a real-time video link supplanted its camera's 2500 ft of film in 1972.

QH-50 DASH

The Navy QH-50 drone evolved from a small, one-man helicopter, the YRON-1 Rotorcycle that the Gyrodyne Company of America had developed in 1955 for the U.S. Marine Corps. Gyrodyne's experience went back to 1946, when it began pursuing the concept of gyrodynes, vertical takeoff and landing (VTOL) vehicles whose engines powered both a rotor for lift and a propeller for forward motion. The Navy requirement for DASH emerged in

Gyrodyne QH-50 DASH.
(Courtesy of the United States Navy.)

the 1950s when the detection range of destroyer sonar systems (22 miles) exceeded the engagement range of their antisubmarine warfare (ASW) weapons (1–5 miles). DASH was conceived to extend the delivery range of ASW torpedoes by launching from the decks of destroyers to prosecute hostile submarines at long range. The Navy awarded the initial DASH contract in 1958.

The QH-50 made its first untethered, unmanned flight on 12 August 1960 at Patuxent River, Maryland, followed by its first flight from a ship at sea that December, when it launched from and recovered back aboard the USS *Hazlewood* (DD 531). It was first deployed operationally aboard the USS *Buck* (DD 761) in January 1963. Over the next 6 years, 771 QH-50C and D models were delivered, and pairs of them were deployed aboard 240 destroyers. Although its primary mission was ASW (it carried two Mark 44 homing torpedoes), DASH was also used in the surveillance, gunfire spotting, bombing, cargo transfer, smokescreen laying, and even rescue roles.

Project Snoopy added a television camera and video data link to selected QH-50s deployed on destroyers off Vietnam to locate targets inland from the coast and adjust 5-inch gunfire against them. DARPA subsequently initiated two projects to improve on this capability. Nite Panther added low-light television, a laser rangefinder and target designator, and an early moving-target indicator (MTI) radar to give it a night/all-weather ability to find targets and direct fire on to them. Nite Gazelle took this capability to the next logical level and added weapons, specifically a minigun, grenade launcher, bomblet dispensers, bombs, and laser-guided rockets. These capabilities were reportedly used effectively against North Vietnamese troops and convoys moving at night.

The DASH established a number of notable firsts for UAVs, to include the following:

1) The first rotary wing UAV produced.
2) The first UAV to take off and land back aboard a vessel at sea.
3) The first unmanned reconnaissance helicopter.
4) The first hunter-killer UAV (Project DESJEZ, employing sonobouys and torpedoes from the same drone). Its fly away price in 1965 was $154,000 ($877,000 in 2002).

Seen as potential competition to the proposed LAMPS ASW helicopter program, the DASH program was terminated by the Navy, who phased out its last system in 1971. A total of 810 had been built of all five models, 786 for the U.S. Navy. Fifty percent (383 of the 771 C/D models) were lost during operations, and 90% of the remainder were expended as target drones through the 1980s. The QH-50 continued to serve in the Japanese Maritime Self Defense Force, who had acquired 24 QH-50Ds, pushing total production to over 800. An employee of Aerodyne Systems attempted to market the QH-50 as the CH-84 Pegasus at the 1984 Paris Air Show, and the German Dornier company's Argus research program used it in the 1980s, leading to the 1990s' Seamos effort to provide a VTOL UAV for the future K130 class of German Navy corvettes. Israel acquired three QH-50s in 1986 and modified them to carry infrared and radar sensors, calling them *Hellstar*. The last 30 active QH-50s are presently employed by the U.S. Army at their White Sands test range to tow targets.

GRD-21

Although the Air Force's Lockheed GRD-21 reconnaissance drone only participated peripherally in the war in Southeast Asia, it possessed performance that remained unmatched for decades in subsequent unmanned aircraft. Designed as an unmanned complement to the Mach 3 A-12 Blackbird reconnaissance aircraft, it was intended to penetrate those environments over hostile territory too dirty (radiation) or too dangerous (SAMs) over which to risk the manned SR-71. Developed under Project

Lockheed M-12 with GRD-21.

Tagboard, it was carried piggyback atop two specially modified Blackbirds (designated M-12s) to altitude, at which point its ramjet was ignited, the drone released, and it flew its preprogrammed 3400-mile route at Mach 3 plus at over 90,000 ft altitude. Once over its recovery zone, its film capsule was ejected for midair recovery and the airframe was expended.

Boeing B-52 with GRD-21s.

The GRD-21's first captive flight occurred in December 1964, followed by its first free flight in March 1966. On its fourth test free flight in July 1966 the drone collided with the M-12 carrier just after its release, destroying itself, the M-12, and killing one of the two crewmen. Tests were suspended until late 1967, when the GRD-21B model emerged, modified to be carried and launched from under the wing of a B-52H bomber. It now relied on a booster rocket to accelerate it to the speed required by its ramjet for ignition. The first B-52 launch took place in November 1967, but the GRD-21 crashed. Additional test flights occurred in 1968.

The GRD-21 flew its first operational mission, over China, in November 1969, but its navigation system apparently malfunctioned, and the drone is believed to have continued on into Siberia. The next two missions successfully returned from over China, but midair retrieval of their film capsules failed, so no intelligence was returned. The fourth and last mission was lost while outbound on its return leg. The program was terminated in 1971 as a concession to China for its help in bringing North Vietnam to the peace talks table. Of the 38 GRD-21s produced, 17 remained in late 1976 when the remaining drones were placed in storage at Davis Monthan AFB, Arizona.

Project Tagboard and the GRD-21 are notable, despite their lack of operational success, because of the performance benchmarks they established for unmanned aviation. Its speed and altitude accomplishments have not yet been officially described, but even so, its admitted speed of over Mach 3 remained unequaled until the NASA X-43A scramjet achieved Mach 7 in March of 2004. As for altitude, the success of the NASA/AeroVironment Helios solar-powered UAV in reaching 96,800 ft in 2002 may have exceeded the achievement of the GRD-21, but with its admitted 90,000-ft+ capability, any claim to that record remains an open question.

If the single largest contribution made by drones during the Vietnam War had to be identified, it would be from the Lightning Bug mission on 13 February 1966. On that flight, a specially modified Bug, equipped with ELINT sensors and a data link to instantaneously relay the sensor data to waiting recorders, flew against a known SA-2 site near Vinh, North Vietnam, on a one-way mission. Its purpose was to lure a SA-2 into firing at it, then collect and relay the electronic parameters of the missile's radio-guidance and fusing systems up to the instant it was destroyed. The mission was successful, and its sacrifice resulted in critical improvements to American electronic countermeasures equipment, enhancing the survivability of manned aircraft for the rest of the war. This one mission was arguably responsible for keeping hundreds of American fighter and bomber airmen from being killed or imprisoned as prisoners of war over the next nine years.

SUMMARY

Wars generally serve as proving grounds for new weapons and tactics, and the Vietnam War certainly served that function for UAVs. The two dozen variants of the Ryan AQM-34 used in that war explored virtually every subtask of intelligence collection, as well as branching into leaflet dropping and chaff dispensing. Their success brought out the "white scarf" mentality for the first time, the opposition of manned pilots to unmanned aircraft encroaching on their role, with which UAVs have since had to contend. Another generalization about war is that periods of disarming and disbanding inevitably follow them. Despite the UAVs record of success over Vietnam (and perhaps somewhat because of it), both the drones and their DC-130 mother aircraft were quickly relegated to aircraft storage facilities (or "boneyards") at the start of the war's drawdown.

Chapter 12

ISRAEL

With 120,000 h of flight time on its UAVs as of 2002, the Israeli Defense Forces (IDF) have one of the thicker logbooks of any nation with operational UAV experience, averaging over 4000 h each year. Israeli UAVs are also among the most widely exported drones in the world, having been sold to over 20 countries. The Israeli aircraft industry began in 1953 as Bedek Aviation, which later evolved into Israeli Aircraft Industries (IAI). Two small Israeli companies dedicated to unmanned aviation, Tadiran and Tamnar, arose in the early 1970s. Tadiran built the Mastiff reconnaissance UAV and Tamnar built target drones and subscale training versions of IAI's UAVs. In 1984, IAI and Tadiran combined to form Mazlat, which is known today as Malat, the UAV division of IAI. Malat is the largest of five UAV manufacturers (the others being Aeronautics Unmanned Systems, Elbit Systems/Silver Arrow, EMIT Aviation Consultants, and BTA Automatic Piloting International) operating in Israel today and holds interests in several of the other companies.

The IDF began considering employing UAVs in 1965 following the deployment of highly capable, Russian-built SA-2 surface-to-air missiles (SAMs) in Egypt and Syria. Three candidates for the reconnaissance mission were examined in light of the new air defense capability, the low and fast Mirage IV, the high and slow U-2, or the unmanned R.20, built by Nord Aviation of France. In March 1967, Israeli officers visited the U.S. to discuss acquiring Ryan Firebee target drones for training purposes, but knowing a variant of the Firebee was flying reconnaissance missions over North Vietnam at that time gave their visit a second level of interest. With no indigenous UAV industry at the time, the IDF was interested only in acquiring a turn-key system, not in attempting to modify a target drone for the reconnaissance role. Visits to Nord Aviation concerning the R.20 occurred in late 1969, but again without a sale.

The downing of two IDF RF-4s by Egyptian SAMs in 1970 sparked renewed interest in acquiring UAV systems. That June, a team of Israeli officers paid a second visit to the U.S., which led to a decision to buy 12 Ryan Firebees modified for reconnaissance. The Ryan Model 124I, referred to as the Mabat ("Observation") by the IDF, began delivery in the summer of 1971.

Its first test/acceptance flight was held over the Sinai Desert on 23 August, and its first operational mission was flown by IDF personnel a few weeks later on 14 September. A reconnaissance version of the Northrop Chukar target drone was also acquired, and both types of UAVs were integrated into the IDF's newly formed 200th Drone Squadron. The squadron saw its first combat missions in the Yom Kippur War in October 1973. The effectiveness of Egyptian SA-2 and SA-6 missiles against Israeli manned fighters during that war increased IDF interest in UAV development and employment, spurring the creation of a UAV division within IAI and the formation of native UAV companies.

IAI's first drones were two decoys, the unpowered UAV-A and the pulsejet-powered UAV-B, which were carried as wing stores on fighters. It also began developing the first of its many twin-boom, pusher propeller designs, the Scout, in 1974 for the reconnaissance role; it entered operational service with the IDF in 1977. Its signature design was inspired by previous IAI success with its twin-boom Arava commuter aircraft, and it had the added benefit of guarding its handlers from inadvertently walking into its spinning propeller when on the ground. Concurrently, Tadiran developed a similar, competing design, the Mastiff. The significant contribution of both decoys and real-time reconnaissance UAVs to the destruction of the Syrian air defenses at the start of the 1982 Lebanon War generated significant interest in the worldwide military community. By the following year, U.S. forces, too, were on the ground in Lebanon and off its coast.

Israeli/Ryan Mabat UAV.

IAI UAV-A Decoy.

The origins of the revival of U.S. interest in UAVs can be traced to the bombing of the Marine barracks in Beirut, Lebanon, in October 1983. As a result of this attack, as well as other hostile actions by various Syrian-supported factions, U.S. policymakers adopted an offensive posture in this region,

IAI Scout UAV.

Tadiran Mastiff MK II.

authorizing bombardment of militia strongholds and Syrian Army positions in
the Shouf Mountains behind Beirut. Both 16-in gunfire from the battleship USS
New Jersey and air strikes by carrier-launched A-6 Intruders and A-7 Corsairs
were employed but with marginal results. Neither the naval gunfire nor the air
strikes were terminally controlled, and prestrike and poststrike reconnaissance
aircraft were routinely engaged by hostile fire. Two U.S. Navy captains, sent
in-country to investigate the shortcomings, came away impressed with the
effective use by the Israeli Army of UAVs in the gunfire spotting role and
recommended the adoption of such a system by the U.S. Navy. John Lehman,
Secretary of the Navy at the time and an A-6 aviator, endorsed this recommendation, and discreet negotiations with Israeli Aircraft Industries and
Tadiran, the Israeli manufacturers of the Scout and the Mastiff UAVs, respec-
tively, resulted in the purchase of a Mastiff system for the Navy.

The Navy took delivery of the system in the summer of 1984 as a "proof of
concept" project, similar to the Advanced Concept Technology Demonstrations (ACTDs) of a decade later. Assigned to the Marine's 1st Remotely
Piloted Vehicle (RPV) Platoon at Camp Lejeune, North Carolina, under
Captain Dee Woodson, USMC, the system was quietly flown in North
Carolina and in Arizona to train its operators and to develop concepts for its
use. The external pilots used for takeoffs and landings were enlisted Marines
who had been identified from their participation in local radio-controlled
(RC) model airplane competitions in Arizona. Five of the top ten pilots

were noted as being Marines; they were subsequently reassigned to the RPV Platoon. Secretary Lehman and then Lieutenant General Al Gray (Commanding General, Fleet Marine Force Atlantic) took personal interests in its progress, dropping in on its training both in North Carolina and in Arizona. Mastiff was unique in that it served as a research and development platform for Naval Air Systems Command while simultaneously supporting operational and exercise commitments for the Atlantic and Pacific fleets.

Led by Major Bruce Brunn, USMC, the 1st RPV Platoon deployed aboard the USS *Tarawa* (LHA-1) in late 1985 for its initial at-sea trials, dubbed Operation Quick Go, to develop a concept of UAV operations in support of amphibious operations. For this operation, field representatives from the AAI Corporation of Hunt Valley, Maryland, fabricated an aircraft arresting hook for the Mastiff so it could trap on the flight deck of the USS *Tarawa*. The unit also improvised a back-up net from a volleyball net in case the hook failed. During their six-month Pacific cruise, Mastiff was subjected to cold weather testing over Adak, Alaska, the jungle climates of the Philippines and Thailand, and the desert environment of Australia. It had logged over 50 h during some 35 sorties by the time the cruise ended in early 1986. Of the five aircraft deployed, the platoon returned with three, the two losses being due to exceeding the allowable design weight with test equipment.

The lessons learned from this deployment led directly to the acquisition of the Pioneer UAV for use by the Navy in a gunfire spotting role later in 1986, a role which later expanded into aerial reconnaissance. U.S. battleships in the Persian Gulf employed their Pioneers with good effect in the fire spotter role during the Gulf War, with Iraqi soldiers surrendering to an orbiting Pioneer in anticipation of their imminent bombardment on one occasion. The association had become Pavlovian; a UAV appeared and a shelling soon followed. Pioneer went on to provide reconnaissance for Marines supporting NATO operations in Bosnia in 1994–1996 and in Kosovo in 1999.

Just as the Mastiff design was a forerunner of that for the Pioneer, so was the *Tarawa* deployment the pathfinder for subsequent UAV deployments aboard battleships and amphibious transports (LPDs). When the last Pioneer eventually retires from U.S. service, it will close a 20-year career of Mastiff/ Pioneer UAVs in maritime use having its roots in the Beirut bombing of 1983 and earlier Israeli success with UAVs in combat.

The IDF found the major shortcoming of its Scout UAVs was the poor reliability of its engine, a German design originally built for motorcycles and later adapted for UAV use. To address this, IAI designed a larger, twin-engine version of Scout called Impact in late 1988. Impact used two of the Scout's engines in a tractor-pusher arrangement; sustained single-engine operation was not possible. It was this design that IAI submitted, under the name Hunter, in response to a 1989 U.S. Army requirement for a "short-range" UAV to

Navy/PUI RQ-2 Pioneer Launch.
(Courtesy United States Navy.)

provide airborne reconnaissance for corps commanders. For Hunter, the twin German motorcycle engines were exchanged first for a pair of English motorcycle engines and then, when they proved inadequate, for a pair of Italian motorcycle engines in an attempt to improve reliability. Single-engine operation was still not possible, however. A fourth engine configuration was explored in 1994, when a Hunter became the first UAV equipped with a heavy fuel turboprop engine. In the Army's 1990 fly-off, Hunter's contractor team of IAI and TRW was selected over the Developmental Sciences/McDonnell Skyeye® competitor. Although some 50 Hunter systems were originally planned to be fielded by the Army and the Navy combined, production ended after only eight systems due to a series of mishaps late in its evaluation phase. Even so, U.S. Army Hunters flew extensively during the 1999 Kosovo conflict and also served with the French and Belgian militaries.

By 1988, the IDF recognized the shortcomings of its Scouts' payload capacity, endurance, and ceiling. IAI developed a larger design with an

Navy/PUI RQ-2 Pioneer Recovery.

improved wing, Searcher, and began flight testing it in 1991, but its engine was still the problem area. An improved variant, Searcher II, changed to a more powerful rotary engine, was flight tested in 1996, and serves today in the IDF.

In the realm of endurance UAVs, the need for a model with even greater payload capacity and altitude and endurance performance led to the Heron design in 1993, which first flew in October 1994. With the higher thrust-to-weight advantage conferred by its Austrian-made rotary engine, Heron has demonstrated an endurance of 51 h and achieved altitudes of 32,000 ft. Using Heron as its entry, IAI again teamed with TRW in 1993 for the American Tier II UAV contract, which the General Atomics Predator®, designed by a former IAI engineer, won. IAI and TRW teamed together for a third effort, an even larger high-altitude endurance UAV entrant in the American Tier II+ competition in 1994, which was won by Teledyne Ryan's Global Hawk design. For this competition, IAI drew on its earlier 1988 design experience with the Hauler, a proposed design for a 50,000-lb, 164-ft wingspan, high-altitude, loitering radar platform. It was developing a similar, conceptual UAV, the HA-10, during the late 1990s as a missile-carrying platform for engaging hostile cruise missiles, which with the Arrow ballistic missile interceptor would form the Israeli equivalent of the U.S. Star Wars effort. In 1999, IAI teamed with the French consortium, EADS, to introduce the Eagle endurance UAV, which was selected by the French military in 2001 for its medium-altitude, long-endurance (MALE) requirement.

IAI also actively tried to introduce UAVs into commercial applications during the 1990s. Its Firebird model demonstrated the utility of UAVs in the wildfire spotting role in an exercise hosted by the U.S. Forest Service in Montana in September 1996. At its conclusion, the forestry community saw no advantage with UAVs in performance or cost over that of manned aircraft currently doing the job.

SUMMARY

With its record of war experience and solid foothold in worldwide markets with its UAVs, Israel ranks as one of the top three nations (with Japan and the U.S.) in UAV production and innovation. Its military forces have employed UAVs in new tactics from general warfare to urban fighting with dramatic results, and its industry has been responsive to the lessons learned in these operations by its military. It was responsible for reintroducing the U.S. Navy to the value of UAVs through the Pioneer program, and its design influence is directly seen in the U.S. Air Force Predator® and U.S. Army Shadow designs. Israeli tactics for employing their UAVs have awakened the world's militaries to their wider potential in warfare, especially as a force multiplier working in cooperation with manned forces, both on the ground and in the air.

Chapter 13

ENDEAVORING TO ENDURE

As the growing sophistication of air defenses pushed aircraft to the limits of aerodynamic flight and the limitations of orbiting reconnaissance satellites became apparent in the 1960s, the need for a new class of reconnaissance capability emerged—above most aircraft and threats but below the low orbiting satellite and with a longer dwell time than either—a pseudosatellite, or "pseudolite." The U-2 had to fly in the "coffin corner" of its flight envelope, one knot from stalling, one knot from overstressing the airframe, and one SA-2 from an international incident. The satellites could glimpse but not watch, and soon those watched knew when to draw the curtains closed. The U.S. Defense Department began pursuing a pseudolite capability based on unmanned aircraft during the latter days of the Vietnam War. They were referred to as high-altitude long-endurance (HALE, or later just HAE) UAVs, where "high" was understood to be above 50,000 ft and "endurance" over 24 h.

COMPASS DWELL

The Air Force began exploring HALE unmanned aircraft for the reconnaissance mission in the late 1960s. It contracted with LTV Electrosystems in 1968 for two optionally piloted (capable of manned or unmanned flight) prototypes, known as the L-450F, which were based on the proven Schweizer SGS 2-32 sailplane. The first flew in February 1970, but crashed (the pilot survived) on its third flight the following month. LTV Electrosystems had meanwhile evolved into E-Systems, and the second prototype had evolved into the unmanned XQM-93 by the time testing resumed. It set the world record for unmanned endurance of some 22 h during a flight in January 1972 at Edwards AFB, California. After the Air Force evaluation was completed in 1972, it was converted back to a manned configuration and used by E-Systems to set 16 international aviation records in altitude and time-to-altitude categories. It was later used by the National Oceanic and Atmospheric Administration (NOAA) for scientific missions.

Thirteen months after the crash of the first LTV prototype, the Air Force awarded a second contract for a competing HALE prototype, apparently as insurance against the failure of the LTV design. The second model evaluated in the Air Force's Compass Dwell program was the Martin Marietta Model 845A, which first flew a year later in April 1972. Two prototypes, both unmanned, were built although parts for a third were also procured. It, too, was based on an in-production Schweizer glider design, the SGS 1-34, making both competitors essentially identical in origin and design and with very similar performance and characteristics. Both even borrowed the same landing gear used by the Grumman Ag-Cat. Its short flight test program, as well as the overall Compass Dwell effort, was concluded that July, and Martin ended by setting a new endurance record for UAVs of 27 h and 54 min. Although neither the XQM-93 nor the Model 845A was selected for production, they established the feasibility of operating UAVs at altitudes above 50,000 ft and for endurances over 24 h.

COMPASS COPE

The Compass Cope program began in July 1971, midway through the Compass Dwell program, with a contract to Boeing for two prototype UAVs, later designated as the YQM-94. The requirements for the two programs were very similar, but with the National Security Agency contributing to Cope's funding, its intent was clearly focused on developing a HALE platform for signals intelligence collection. Mirroring the history of the Dwell effort, a competing Cope award was made well after the first was

Air Force/LTV XQM-93.
(Courtesy United States Air Force Museum.)

Air Force/Martin Model 845A.
(Courtesy United States Air Force Museum.)

underway. The Boeing YQM-94 Gull, or Cope-B (for Boeing), first flew in July 1973 and, unlike the first Dwell prototype, crashed on its second instead of its third flight that August. The 15-month hiatus until Boeing's second prototype was ready to fly allowed the competing Teledyne Ryan YQM-98 Tern, also know as the Cope-R (for Ryan), to catch up. It first flew in August 1974, three months before the resumption of Boeing testing. On its fifth and final evaluation flight at Edwards AFB that November, it set a new endurance record of 28 h and 11 min, beating the Dwell record by 17 min. The evaluation of the Cope-B ended that same month, with the Cope-B demonstrating an endurance of just over 17 h. Cope-R was then airlifted to Patrick AFB, Florida, where it underwent a second round of evaluation during 12 flights from May to September 1975, one aircraft being lost during a night landing. The following August the Air Force selected Boeing to build three preproduction Cope aircraft, but then terminated the preproduction contract in July 1977 due to the technological immaturity of the Precision Location Strike System, for which Cope was to have acted as a relay.

Teledyne Ryan's Cope-R design, the Model 275, owed much to the company's previous Compass Arrow (AQM-91 Firefly) reconnaissance UAV program and foreshadowed many external features to be used in its Global Hawk design nearly 20 years later. Both Dwell and Cope approached the development of their HALE UAVs by modifying conventional aircraft/glider designs to fly between 50,000 and 60,000 ft for up to a day. Their accomplishments are summarized in Table 13-1.

Table 13-1   **Summary of Compass Dwell and Compass Cope programs**

	Compass Dwell		Compass Cope	
Designation(s)	XQM-93A L-450F	— Model 845A	YQM-94A Cope-B "Gull"	YQM-98A Cope-R "Tern"
Manufacturer	LTV (E-Systems)	Martin Marietta	Boeing	Teledyne Ryan
Contract Award	1969	Apr 1971	Jul 1971	Jun 1972
Number Built	2	2	2	2
First Flight	Feb 1970	Apr 1972	Jul 1973	Aug 1974
Max Altitude (demonstrated)	50,000+ ft	40,000+ ft	55,000+ ft	55,000+ ft
Max Endurance (demonstrated)	22+ h	27 h 54 m	17 h 24 m	28 h 11 m
Powerplant	Turboprop	Recip	Turbojet	Turbojet

DARPA PROJECTS

Endurance UAV development shifted to DARPA in the early 1980s, and new technologies for sustained unmanned flight, such as fully automated flight control and solar electric propulsion, were explored. Three long-endurance UAV programs began during the decade: HALSOL in 1983, Condor in 1984, and Amber in 1984. The latter two were developed under DARPA's Teal Rain umbrella program.

Air Force/Boeing YQM-94 Gull (Cope-B).
(Courtesy United States Air Force Museum.)

Air Force/Teledyne Ryan YQM-98 Tern (Cope-R).

Under a classified program, AeroVironment developed and flew the High-Altitude Solar Powered (HALSOL) flying wing in 1983, which although it never graduated from battery to solar power, did serve as the forerunner to a series of highly successful sun-powered UAVs under NASA sponsorship in the 1990s (see Chapter 14). It was an early effort to power an aircraft fully or at least partially by solar cells.

DARPA followed with Boeing's twin-engined Condor, a huge (201-ft wingspan) twin-engined HALE UAV that pioneered a number of innovations for future UAVs. Condor was the first aircraft to make a fully autonomous flight, including an automated takeoff and landing, and the first to have automated failure management, allowing it to recover from certain in-flight emergencies, such as engine failure. Arriving at Boeing's Moses Lake, Washington, flight test airfield in March 1986, the two prototypes set a number of unofficial world records during their next 2 years of flying, including 67,028 ft for altitude by a piston-powered aircraft and 51 h at 55,000 ft for endurance by an unmanned, unrefueled aircraft. It made a total of eight flights. The Navy considered Condor for the broad ocean surveillance and communication relay missions but did not pursue this application. The surviving Condor now hangs in the Hiller Aviation Museum in San Carlos, California.

Amber was a joint DARPA/Navy effort to build a low-cost, medium-altitude, long-endurance (MALE) UAV capable of being used either as a weapon or for long-term surveillance. The design was originated by

Abraham Karem, an expatriate Israeli and former Israeli Aircraft Industries engineer who had founded his own company, Leading Systems, Inc., to pursue his approach to endurance flight. Receiving $40 million from DARPA in December 1984, Leading Edge built six Amber prototypes over the next 2 years with Karem's signature inverted V-tail. Three of them, with pointed noses, were strike variants, intended to deliver a weapon by jettisoning its wing and falling on its target, just like the Curtiss-Sperry Aerial Torpedo and Kettering Liberty Eagle of World War I. The other three, with rounded noses, were to carry television cameras and a high-bandwidth data link for flying real-time video surveillance missions, referred to as a "CNN in the sky" capability. Amber was flight tested at El Mirage in the California Mojave Desert from November 1986 through June 1987. It then relocated to the Dugway Proving Grounds in Utah for follow-on Army tests that fall, demonstrating an endurance of 35 h and an altitude of 27,800 ft. In a public display of its capabilities, Amber's progress toward an endurance record of 38 h and 22 min was displayed in real time at an exhibit at the Association of Unmanned Vehicles International 1988 convention. A third evaluation was conducted in 1989, this one by the Army at Fort Huachuca, Arizona, and focused on its operational suitability. Twelve Ambers (one had crashed) were put into storage when the project ended in 1990.

When Amber development was over its peak, Leading Systems decided to build a low-cost, export version of it called the Gnat-750™. Its first flight occurred in mid-1989, but Leading Systems was facing mounting financial

DARPA/Boeing Condor.

DARPA/Leading Systems Amber.

troubles, which led to its being bought by General Atomics in 1990. General Atomics' expertise was in the nuclear reactor business, so it established a new subsidiary, General Atomics Aeronautical Systems, Inc., dedicated to developing UAVs. Testing of the Gnat continued, steadily building its endurance up to 40 h, and by 1993, it had attracted two customers, the Turkish government, who ordered a system with six aircraft, and the U.S.'s Central Intelligence Agency (CIA). The CIA was looking for a low-cost, low-risk way to monitor the Bosnian conflict and support U.N. peacekeeping operations there. The operational concept was to have the Gnat loiter quietly over trouble spots, relay its electro-optical visual and infrared video through a manned Schweizer TG-8 airborne relay aircraft to its forward ground station, who would then beam the Gnat's video via satellite in real time (minus the few seconds of analog to digital conversions and transmission lags) to watchers in Washington, D.C. When it was found the Gnat's data links would interfere with Italian television broadcasts, its planned forward base of operations was moved from Italy to Albania, a country now anxious for Western ties after its recent emergence from communism. The Gnat-750™ team deployed there in February—March 1994, then returned the following winter. This time it was based on the Croatian coast, to which it redeployed for two to three months at a time over the next several years. Five such combat deployments had been completed by July 1996. While the Gnat's video proved effective when it could get into position, the harsh Balkan weather severely limited these opportunities.

In an effort to consolidate the U.S. government's proliferating UAV projects, the Congress created a Navy-led Joint Program Office (JPO) for UAVs in 1989 to provide program management and subsequently established the Defense Airborne Reconnaissance Office (DARO) in 1993 to oversee their acquisition. Their strategic game plan envisioned three levels, or tiers, of UAVs, Tier I for a tactical endurance UAV, Tier II for a theater (operational level) endurance UAV, and Tier III for a strategic endurance UAV, a national-level intelligence asset much like the U-2. The Gnat-750^{TM} was identified as the already-in-being Tier I asset, but systems to meet the Tier II and Tier III requirements had to be found.

ADVANCED CONCEPT TECHNOLOGY DEMONSTRATION (ACTD)-BORN UAVS

At this same time (1993–1994), the Pentagon acquisition leadership introduced the concept of ACTDs. ACTDs were intended to apply mature technologies in innovative ways and rapidly put the capability in the hands of overseas theater commanders for their field evaluation in the "real world." The entire process from start to finish was to take no more than 3 to 4 years, which would avoid the traditional acquisition process that took 10 to 20 years from concept to delivery, and would allow for inputs from the ultimate customers, the warfighters in the theaters; i.e., to be joint rather than Service-centric in perspective.

The first class of ACTDs included both the Tier II and Tier III UAV projects, and indeed, over the coming years, ACTDs became the preferred route for acquiring Defense Department UAVs. They were bare-bones funded to quickly produce one or two examples of the concept without long-term industrial commitments, possibly because the theater commanders could judge them to be of no real value. They were experiments in which there were failures and where rejection still represented success. The problem was that no one had mapped out how to transition an ACTD product that the theater heartily endorsed into a producible, sustainable system, complete with a training syllabus, a spares inventory, or a logistics chain to support it.

It was also an uninvited guest. The bureaucracy that produces the defense budget is a rolling process with a long outlook (7 years ahead) and a lot of inertia; old programs are hard to kill and new ones hard to start. Inserting change in this process has been compared to turning the Titanic. ACTDs would appear suddenly in this measured dance, demanding funds where none had been allocated, which meant existing programs often had to be bled to pay the ACTD's bill. Although the process for transitioning ACTDs to formal acquisition programs has been smoothed out in recent years by adding a mandatory transition phase, using the "A-word" around service budget officers even today will evoke a reaction similar to that from saying "bomb" in an airport.

General Atomics ASI already had its design for a larger version of the Gnat-750™, called Predator®, on the drawing board when the Tier II requirement came out in a memo from John Deutsch, then Undersecretary of Defense for Acquisition and Technology, in July 1993. It called for a UAV capable of loitering for 24 hours 500 miles from its base and carrying a 400–500 payload that could provide images with 1-ft resolution. General Atomics ASI was put on a $32 million Navy contract for Tier II on 7 January 1994, and Predator® made its first flight, a 20-second hop at El Mirage, California, on 3 July 1994, just in time to meet the contract's 6-month requirement. That January, Predator® demonstrated an endurance of 40 h and 17 min at Fort Huachuca before participating in the joint Roving Sands military exercise held in New Mexico that spring, during which it flew on 25 of the exercise's 26 days, imaged over 200 targets, and provided 85% of the imagery collected during it.

Coming off this impressive showing, the Predator® ACTD team met with U.S. European Command (EUCOM) officers that May to discuss using Predator® in the real-world contingency over Bosnia. Up to this point, Predator® had been operated in the conventional fly-to-a-point, take-a-picture, fly-to-the-next-point mode of previous intelligence collectors, despite being equipped with a video camera in a gimbaled turret. The EUCOM officers were now used to receiving video in real time from the CIA Gnat-750™ deployed in their theater and expected Predator® to provide the same. The paradigm was changed, and when Predator® arrived in theater that July, it too provided video surveillance instead of snapshot reconnaissance. Its original 60-day stay (Operation Nomad Vigil at Gjader, Albania) was extended to 120 days, and it was called back to help enforce the Dayton Peace Accords for Bosnia in March 1996 (Operation Nomad Endeavor), this time flying from Taszar Air Base in southern Hungary. In the middle of these operations, Washington decided that Predator® had demonstrated military value (June 1996) and approved its transition to a formal acquisition program (August 1997), making it the first ACTD to "graduate." With one brief respite during the winter of 1997–1998, at least one Predator® system has been continuously deployed overseas since that time.

One of the lessons learned during the 1999 Kosovo conflict was that the lag between spotting a target and attacking it, known in military circles as the sensor-to-shooter kill chain, was unacceptably long, allowing too many found targets to escape. By one account, of the nearly 300 Serbian vehicles located by Predator® and Hunter in Kosovo, only three were subsequently attacked by fighters on call for the task. To shorten the kill chain, the Air

Force replaced its standard gimbaled camera turret on Predator® with one containing a laser designator for targeting. It then added Army antitank AGM-114 Hellfire missiles, enabling one Predator® to find, designate, and attack these moving targets. The combination was tested in February 2001 at China Lake, California, and by October was being used in Afghanistan and later in Kuwait against terrorists' vehicles. Although not designed to be stealthy, the combination of Predator's® thin silhouette, fiberglass airframe, low speed, and altitude make it virtually invisible to opponents without an integrated air defense.

DARKSTAR AND GLOBAL HAWK

When the original definition of Tier III as a large, ultralong range, stealth, penetrating UAV was developed in 1993, it was made to mirror an ongoing special access Air Force program. Similar in size and price tag to the Northrop B-2 manned bomber, Lockheed Martin Skunk Works' Quartz UAV was canceled soon after its Cold War justification evaporated, but not before $1 billion had been spent on it. Quietly, however, Lockheed continued working on a scaled-down version of it, called DarkStar.

Meanwhile, DARPA keyed on shortfalls in intelligence capabilities identified from the Gulf War and its previous experience with HALSOL, Condor, and other programs to initiate the High-Altitude Endurance (HAE) reconnaissance UAV program in April 1994 as a way to meet some of the requirements for the Tier III UAV. When the Office of the Secretary of Defense (OSD) "invented" ACTDs later that same year, the HAE UAV

CIA/General Atomics Gnat-750™.

USAF/General Atomics RQ-1/Predator®.

program was designated as one of the initial ACTDs. Five designs, from Loral, Northrop Grumman, Orbital Sciences, Raytheon, and Teledyne Ryan Aeronautical (TRA), were initially evaluated, with the TRA Global Hawk design being selected in May 1995. But Global Hawk was a conventional aircraft, designed to stand off from hostile airspace while collecting intelligence. Lacking the stealthy penetrating capability envisioned for Tier III, it was labeled "Tier II+." To fulfill the Tier III requirement, it was combined in 1995 with the previously under development DarkStar stealth reconnaissance UAV into one joint HAE UAV program. This dual aircraft ACTD effort was to provide both a conventional design, Global Hawk (also known as Tier II+ or CON HAE), for use over low- to moderate-threat environments and a complementary low observable design, DarkStar (also known as Tier III- or LO HAE), for penetrating high-threat areas.

DarkStar flew first on 29 March 1996, but the one prototype crashed on its second flight three weeks later. Analysis of the mishap took 27 months and revealed a number of engineering shortfalls that had to be designed out of subsequent models before flying could resume. The second prototype took to the air in June 1998. The Defense Department canceled the DarkStar portion of the ACTD in January 1999, just before delivery of the third prototype.

One of the contributors to DarkStar's demise was the large difference in its performance compared to its ACTD stable mate, Global Hawk, and this comparison provides an interesting analysis of the cost of stealth. Because both UAV programs worked to the same, single requirement, a $10-million unit flyaway price (in 1994 dollars, or $12 million in 2002 dollars), their performance was allowed to float to meet this cost constraint. With the only added requirement being that of stealth, Lockheed was constrained to produce an aircraft one-third the size (8600 lb) and with one-third the performance (12 h endurance) of the nonstealthy Global Hawk. Simplistic though it may be, this comparison illustrates the high cost paid for the leverage stealth technology provides. While such an investment is unquestioned for manned aircraft where stealth helps ensure aircrew survival, its merit for use with unmanned aircraft is debatable.

One year after its roll-out ceremony, Global Hawk made a successful 57-min first flight at Edwards AFB, California, on 28 February 1998. Over the next 28 months, the 5 Global Hawk test vehicles demonstrated repeatable, reliable performance, reaching an altitude of 66,400 ft and an endurance of 31.5 h in the course of completing its ACTD. More important for future UAV operations, it flew over half of its hours in civilian-controlled airspace (outside of military ranges), setting precedents with the FAA and European airspace authorities for approving future HAE UAV flights.

DARPA transferred program management responsibility to the Air Force midway through flight testing in October 1998. During the course of the last phase of its ACTD, the Military Utility Assessment (MUA), Global Hawk flew successively from California to Alaska, across the United States, then across the Atlantic to Portugal and back, nonstop, making the first transoceanic round trip by a UAV in May 2000. (The first one-way trans-Atlantic UAV flight had been made in August 1998, by the Insitu Group's 31-lb Aerosonde™.) Its test program, while free of fatalities, was marred by two mishaps: the first when its flight termination sequence was unintentionally activated at 41,000 ft, resulting in the aircraft spiraling into the Mojave Desert, and the second when a software anomaly accelerated the aircraft at high speed off the Edwards AFB taxiway during ground operations, damaging a second aircraft. Upon the ACTD's conclusion in June 2000, Joint Forces Command (JFCOM) evaluators found Global Hawk had successfully demonstrated its military effectiveness, suitability, and inter-operability and recommended it for production, referring to it as the "theater commander's low-hanging satellite." It won the 2000 Collier Trophy for aviation achievement, the first UAV to do so.

On 22–23 April 2001, Global Hawk became the first UAV to cross the Pacific, flying from Edwards AFB, California, to RAF Edinburgh, Australia, outside Adelaide, covering the 7267-mile route in 23 h. Named *Southern Cross II* in honor of the first manned aircraft to cross the Pacific, it flew 12 missions in and around Australia over the next 6 weeks before retracing its route home on 7 June. In November 2001, Global Hawks were pressed into service flying combat missions over Afghanistan in support of the war on terrorism and, in 2003, over Iraq. The Air Force plans to formally start

DARPA/Lockheed RQ-3 DarkStar.

operations in 2006 with the first of a planned fleet of 51 Global Hawks to be stationed at Beale AFB, California.

AEROSONDE™

Seventy-one years after Lindbergh's solo crossing of the Atlantic, a small, unmanned aircraft replicated his feat. Doctors Greg Holland and Tad McGeer of The Insitu Group, a small firm located in the Columbia River gorge of the Pacific Northwest, had begun developing the Aerosonde™ in 1992 as a small- (30-lb) but long-endurance (30-h) carrier for meteorological instruments into open-ocean regions seldom monitored for weather. Designed to be a totally automatic plane, it borrowed a page from Lawrence Sperry's book on launching UAVs by riding in a special car roof rack until the car reached flying speed. At that point, about 50 mph, the driver would reach out, trip a release, and the drone would lift up and away. At mission's end, it would skid to a stop on its reinforced belly. Elegant, reliable, and inexpensive, these techniques sidestepped what had been a nearly insurmountable hurdle for Sperry, Kettering, and the other early pioneers, and allowed Insitu to focus on performing the airborne mission.

Good endurance was critical to competing with weather balloons, and by 1996, they had flown Aerosonde™ on a 24-h flight. Its small size precluded

Air Force/Northrop Grumman RQ-4 Global Hawk.

carrying the bulk of a satellite data link for communicating while over the horizon from its ground station, which was for most of its flight, so it had to be capable of thinking for itself during long periods of isolation. By early 1998, they had it execute a completely autonomous flight, including its takeoff and landing. With these capabilities demonstrated and over 800 h of flight experience with the Aerosonde™ under their belts, the team was now ready to attempt the first transoceanic flight by an unmanned aircraft.

In August 1998 a joint Insitu/Washington State University team arrived with four Aerosondes™ in crates at Bell Island Airport off the coast of Newfoundland, Canada. Another portion of the team deployed to a British military test range at Benbecula on an island in the Outer Hebrides, off the coast of Scotland. The team's strategy was to wait until the right North Atlantic frontal systems were in place to provide a tailwind for the entire west-to-east flight before launching. On 17 August, the first Aerosonde™, *Piper*, was chauffeured into the air only to crash minutes later when control was switched from manual to autonomous. A software error was promptly identified as the culprit and fixed; Then *Trumper* was launched, but it never arrived over Scotland, lost somewhere over the North Atlantic while out of link range. *Millionaire* quickly followed and just as quickly disappeared too.

Launching *Laima* from Bell Island.

At dawn on 20 August, the fourth and last Aerosonde™ was mounted atop the rental car, driven down the airport's runway, and released into the air.

For the next 26 h and 45 min (coincidentally, close to Lindbergh's flight duration), *Laima*, named for the Latvian goddess of good fortune, skirted along the edge of two frontal systems at 5500 ft recording its location, altitude, and the winds it encountered. For nearly 18 h it flew through moderate to heavy rain showers. Just after noon the next day at Benbecula, the forward team reestablished contact with *Laima* and manually guided it into a landing on the Scottish heather. *Laima* had not only become the first UAV to cross the Atlantic but also the smallest airplane ever to do so. When the team opened the drone up, water from the storms it had weathered was found sloshing around inside the small fuselage. The entire 2031-mile flight had been flown on one and a half gallons of fuel. Today, *Laima* hangs, suspended in flight, in Seattle's Museum of Flight.

SUMMARY

Man's presence in the air has been gradually extended through a series of technological advancements and new piloting skills. When the amount of fuel carried was the original limiting factor, aviators developed the technique of aerial refueling, tentatively in the 1920s and then with ubiquity by the 1950s. The next barrier to long flight, oil consumption, fell away as reciprocating engines gave way to turbine engines. The first nonstop, around-the-world flight by the B-50 (Lucky Lady II) in 1949 (with extra crewmen onboard) capitalized on the downfall of this barrier while also making a pointed demonstration that anyone was now reachable in one flight. The final barrier became man himself, both psychologically to long-duration flights and physiologically to high-altitude missions. When new missions, military and scientific, emerged that dictated aircraft fly even higher and longer, man, the aviator, began relinquishing his role to robotic aircraft.

Endurance UAVs introduced not just new technologies, such as solar-powered flight, but new operational paradigms as well, such as the concept of a pseudosatellite (or "eternal airplane"). Predator® and Gnat™ not only introduced a new capability (real time surveillance) to military commanders but also served as ambassadors to former communist bloc countries (Albania, Croatia, Hungary), whose introduction to military cooperation with the United States after the fall of the Berlin Wall was via these early endurance UAV deployments.

Chapter 14

Unfueled and Unmanned

The Sun continuously rains energy at the rate of 100 W per ft^2 on the Earth. A modest size aircraft typically has a wing area of 100 ft^2. Such aircraft require about 250–500 W for each knot of airspeed flown.

In the early 1970s, Robert Boucher put these three facts together and decided to build a solar-powered aircraft using the relatively new technology of photovoltaic (PV) cells, then used to power satellites and space probes. He imagined an aircraft with the ultimate endurance—unlimited. In 1974, Lockheed, working under a DARPA contract, gave him a subcontract to turn that concept into reality. Within months, with the help of his twin brother Roland, he had built the *Sunrise I*, the world's first fuelless-powered airplane. Its 32-ft wing had a large patch of 1000 solar cells across its upper surface that could provide up to 450 W. PV cells of that day only had efficiencies of 10–15%, were inflexible, and were expensive. The 26-lb *Sunrise I* made the world's first solar-powered flight from Bicycle Lake, a dry lakebed in California's Mojave Desert, on 4 November 1974. Its PV cells were the sole power source for its two ferrite electric motors; no backup batteries were carried. It made a number of successful flights before being damaged in a wind storm that spring.

Boucher received an immediate follow-on contract to build a second solar airplane in June, which was ready for flight within three months. On 27 September 1975, the *Sunrise II* took to the air at Nellis Air Force Base, Nevada. Its 90 ft^2 of wing surface contained 4480 PV cells that generated 600 W for its single cobalt electric motor. These cells weighed 4 lb compared to the Sunrise II's 22.5. It demonstrated a climb rate of 300 ft/min and an altitude, limited by its data link, of 20,000 ft. Its calculated performance was estimated to be 75,000 ft in summer and 45,000 ft in winter, when Sun angles were more oblique.

Boucher documented his experience developing solar-powered flight in a 1978 book, *The Quiet Revolution*, which captured the attention of Dr. Paul MacCready of AeroVironment, Inc., who was pursuing human-powered flight at the time. MacCready teamed with Boucher in 1979 and, with funding from DuPont, built the world's first man-carrying solar airplane, the *Gossamer Penguin*. Using the 600 W panel of PV cells from the *Sunrise II*, mounted umbrella fashion over its wing, and Boucher's cobalt electric motor, the *Gossamer Penguin*, with Janice Brown as its pilot, made the world's first manned solar-powered flight on 16 May 1980 at an abandoned airstrip near Shafter, California. MacCready and Boucher next teamed on Dupont's *Solar Challenger*, whose 16,128 PV cells delivered 2600 W. On 7 July 1981, Steve Ptacek piloted the 200-lb *Solar Challenger* from Pointoise, France, to RAF Manston, England, becoming the first solar-powered airplane to cross the English Channel, 72 years after Louis Bleriot's historic first Channel crossing.

Building on their 1974–1975 work with Boucher, Lockheed won three successive contracts from NASA Langley in 1981–1984 to do a concept design for its Solar High Altitude Powered Platform (Solar HAPP) project. Designed to fly for up to a year at 100,000 ft, the Lockheed Solar Star's mission was intended to be precision agriculture, monitoring crops in California's Central Valley with infrared and ultraviolet imaging sensors to detect early signs of crop disease, water stress, fertilization level, and other factors impacting crop yield. The Solar Star design had several innovative features. First it was to be towed into the air by an airplane, then released to spiral up to its operating altitude. Second, its outer wing panels were to have PV cells on their upper and lower surfaces, and these panels would rotate into the vertical position during the day to present more area to sunlight, then fold down horizontally at night to optimize cruising. Third, the excess solar energy collected during the day was to separate water carried onboard into hydrogen and oxygen, which would feed a fuel cell for aircraft power through each night. This concept for using a solar-fuel cell cycle for day/night operation would resurface 20 years later in the NASA/AeroVironment effort to fly Helios over multiple days.

Boucher's future efforts focused on improving electric motor performance, but AeroVironment and MacCready remained interested in solar-powered flight. The Gossamer Penguin and Solar Challenger had been extremely fragile structurally and their small wing area had limited their power draw from sunlight. Ray Morgan, a coworker of MacCready's, saw a solution in a carbon and Kevlar airframe with Mylar plastic covering a much larger wing, one that made the design a virtual flying wing. With funding from a special access Defense Department program, AeroVironment built the High-Altitude Solar (HALSOL) unmanned airplane and flew it at Groom Lake, Nevada, in June 1983. Although it carried a panel of solar cells, their purpose was to evaluate the effect of wing flexure on their performance; HALSOL was actually powered by batteries. In a series of test flights that summer, the 410-lb plane reached 8000 ft and flew for an hour, but the end conclusion was that the PV cell efficiency of that day was insufficient for sustaining long-endurance, high-altitude solar flight.

DARPA/Astro Flight Sunrise I.

A decade later, PV cell efficiencies had reached 19%, and there was renewed interested in HALSOL from the Ballistic Missile Defense Organization (BMDO), who were interested in its potential ability to carry sensors for detecting SCUD missile launches, a requirement that emerged from the recent Persian Gulf War experience. HALSOL's variable pitch propellers were replaced with fixed pitch ones, its DC electric motors with brushless AC ones, and its PV cells with more, higher efficiency versions. Pathfinder, as it was now called, now carried 200 ft^2 of PV cells on its 600 ft^2 wing, and they provided 3800 W. The unmanned Pathfinder made its maiden flight on 20 October 1993 from the dry lakebed at Edwards AFB, California, flying for

SOLAR HAPP

NASA/Lockheed Solar HAPP proposal.

41 min and reaching 200 ft. It was 60% solar- and 40% battery-powered for its series of flights that fall.

Late that year, NASA assumed sponsorship from BMDO, and Pathfinder joined the stable of UAVs NASA was assembling for its Environmental Research Aircraft and Sensor Technology (ERAST) program to evaluate UAV suitability for conducting high-altitude, long-endurance (HALE) science missions. Such missions would be useful for conducting both "process" studies—observations of long-term phenomena, such as the ozone hole, tied to diurnal or even seasonal cycles—as well as "cal/val" purposes—calibrating satellite instruments and validating their data by making coordinated observations simultaneously with their orbiting counterparts. NASA was trying to find a platform capable of carrying scientific instruments for sustained observations in the no-man's zone between aircraft and satellite coverage shown in the figure below. This is the flight regime where altitude requires man to wear a full pressure suit and endurance pushes him beyond the boundaries of human fatigue, either consideration making man the limiting factor to flight in this zone.

Managed by NASA's Dryden Flight Research Center, the ERAST program selected four UAVs to prove their payload and endurance capa-

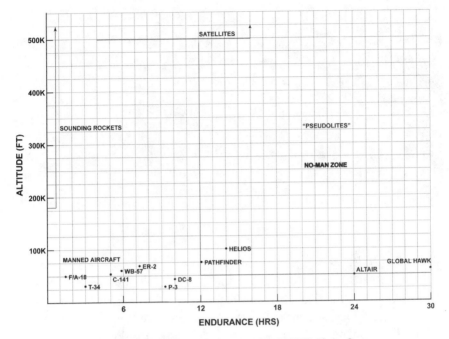

Altitude-endurance comparisons for HALE Aircraft.

NASA/AeroVironment Pathfinder.

AeroVironment HALSOL.

bilities in a semicompetitive atmosphere; varying degrees of success were achieved. Scaled Composite's D-2 Raptor demonstrated over-the-horizon data linking via the Tracking and Data Relay Satellite System (TDRSS) in 1996, but the one aircraft was later destroyed in a crash. Scaled Composites' manned Proteus aircraft subsequently joined the ERAST program in its latter stages. Aurora's Perseus tested ways to enhance the performance of a conventional internal combustion engine for high-altitude flight, setting an unofficial record for turbocharged, single-engine, propeller-driven aircraft of over 60,000 ft. General Atomics' Altus®, a variant of the Air Force's RQ-1 Predator® reconnaissance UAV, stressed endurance at more modest altitudes, flying for 26 h carrying an atmospheric radiation measurement experiment, although it demonstrated 4 h endurance above 55,000 ft with turbocharging. The fourth was AeroVironment's Pathfinder.

Under ERAST, PV cells were incrementally added to Pathfinder's upper wing, covering 70% of it initially and eventually 100%. Because sun angle and length of day were critical at this early stage of solar-powered flight, Pathfinder's flying season was limited to the spring and summer months. It began its first season of flight tests at NASA Dryden in April 1995, concluding it with a record-setting flight to 50,500 ft on 11 September 1995. A month later, it was severely damaged while parked in a hangar, and its 1996 flying season was lost to repairs. For its second flying season, in 1997, Pathfinder was relocated to Kauai, Hawaii, where in a series of 7 flights, it carried a camera to image Kauai's coral ecosystem and bettered its own altitude record, reaching 71,530 ft on 7 July 1997. Its mission profile was to launch in the still predawn air, spend the day climbing under solar power, then rely on battery power and its 30:1 glide ratio to descend and land that night.

During the winter of 1997–1998, the center section of Pathfinder was replaced with a larger section, increasing its span from 98 to 121 ft and adding two more electric motors and propellers for a total of 8. This new configuration was labeled Pathfinder Plus. During the 1998 flying season at Kauai, it again raised the altitude record for both propeller-driven and solar-powered aircraft on 6 August 1998 to 80,201 ft.

While Pathfinder Plus was deployed to Hawaii, its successor, Centurion, was being built with an even stronger composite structure. Weighing 1235 lb and mounting 14 electric motors and propellers on its 206-ft wingspan, it was to be capable of carrying a 600-lb payload to 100,000 ft. Delivered to Dryden in September 1998, Centurion made the first of its three test flights, all under battery power, on 10 November 1998, and its last, at its maximum flying weight of 1806 lb, on 3 December. It had been planned to evolve Centurion's design to the slightly larger Centelios configuration, then to the ultimate Helios design, but ERAST out-year funding was considered more of a risk to the program than technology at that point, so Centurion's transformation directly into Helios began that January. There was also a debate as to whether to continue to use lower cost, lower efficiency PV cells or convert to higher efficiency, much higher cost ones.

Helios was designed to achieve two goals established for the ERAST program: flight to 100,000 ft and flight above 50,000 ft lasting 4 days (96 h) by 2005. Its initial battery-powered flights were made in September 1999. Its wing now spanned 247 ft and was covered with 1500 ft^2 of 62,130 PV cells capable of generating 20 W/ft^2. Its 14 2-hp electric motors consumed 21 of

Table 14-1   Evolution of the solar-powered UAV

Selected data	Aircraft				
	Sunrise I	Pathfinder	Pathfinder Plus	Centurion	Helios
First flight	4 Nov 74	20 Oct 93	98	10 Nov 98	8 Sep
Gross weight, lb	27	410	700	1806	2048
Payload weight, lb	0	40	150	571	726
Motors/propellers	2/1	6/6	8/8	14/14	14/14
Wingspan, ft	32	98.4	121	206	247
Wing area, sq ft	90	787	968	1648	1976
PV array size, sq ft	72 (est)	550	658	Not installed	1580
PV array output, kW	0.45	3.8	12.5	Not installed	30
Altitude achieved, ft	20,000 (Sunrise II)	71,530	80,201	<1000	96,863

the 30 kW generated by this solar array. The 2000 flying season was spent installing the more capable solar array. Returning to Kauai the following summer, Helios again established a new altitude record, 96,863 ft, for sustained level flight on its 13 August 2001 sortie. The 2002 season was used to fly two experiments while its day/night power system was developed: one a Japanese payload for relaying high-definition television (HDTV) and the other a set of cameras for assessing the ripeness of coffee beans synoptically over the large area of a nearby Kauai coffee plantation. Eventually, Helios's regenerative energy storage system, in which its solar array decomposes water into hydrogen and oxygen by day which then serve as fuel for its proton exchange membrane (PEM) fuel cell will be used to power the aircraft through the night perhaps for months at a time.

The ERAST program produced three positive outcomes for UAVs by 2002. Based on Altus'® performance, NASA selected (and helped fund) the General Atomics' Predator B® as the winner of the ERA (Earth Research Aircraft) portion of ERAST for future UAV-based studies. Second, NASA established a dedicated solar-powered aircraft (SPA) effort to evolve the AeroVironment Pathfinder design to one called Helios, which was to be capable of flying to 100,000 ft or remaining above 50,000 ft for 4 days. Key to AeroVironment's "eternal airplane" effort was development of a regenerative solar energy storage system to store solar energy overnight and thereby enable sustained flights of a month or more. Third, NASA was able to demonstrate that useful science could be done from HALE UAVs by assessing Hawaiian coral reef erosion and coffee crop ripeness, fulfilling ERAST's original goal of a decade earlier of regularly doing science in the no-man's zone between conventional aircraft and satellites.

SUMMARY

In terms of capability, flying unfettered from the limitations of carrying fuel onboard has a special appeal to environmentalists and presents an attractive option for science and military mission planners. Loitering uninterrupted at great altitude over days or months would allow unprecedented access to such phenomena as ozone holes and polar ice melting, as well as providing a stationary perch short of geosynchronous orbit for early warning sensors. Great progress has been made in the three decades of solar-powered flight, largely attributable to the focused effort in this field brought to bear by NASA and AeroVironment over the course of the past decade (see Table 14-1). In terms of technology, despite payload fractions reaching a respectable 35%, ultra-low wing loadings of 1 psf continue to dictate large, flimsy structures to carry a worthwhile amount of payload. Structural strength is therefore the limiting factor for payload capacity in these designs, and that limit is essentially reached with currently available materials. Surviving through the night, repeatedly, is the next major challenge to the acceptance of solar-powered UAVs in commercial, scientific, and military roles. More efficient, less expensive PV cells are key to enabling both.

Chapter 15

UAVs Today: A Snapshot

What is the result of the history of unmanned aviation that has been described up to this point? This chapter and the following depart from the historical nature of the preceding chapters to describe the present state of UAVs to which this history has led and to forecast where it may lead in the future.

Worldwide UAV Census

Today, there are an estimated 2400 UAVs—military and civilian—in use around the world. The overwhelming majority, 65% of them, are Japanese-produced, radio-controlled helicopters used for agricultural purposes. Japan has 1565 such UAVs and 6000 licensed operators for them, and they perform about one-third of the agricultural aviation in Japan, covering some 10% of their total rice acreage. Commercial sales of them amount to about $100 million annually. The Japanese UAV market began with a government-sponsored competition in 1986 to find a way to compensate for the dwindling population of rice farmers, an occupation in decline as the children of this traditional way of life opted instead for urban jobs. Robotic helicopters were selected, and the government subsidized their initial development. By one estimate, each robotic helicopter can do the work of 15 farm laborers.

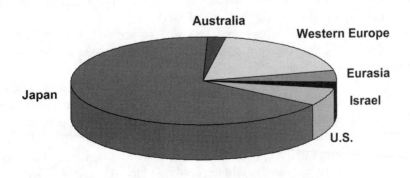

Proportions of the 2400 UAVs employed worldwide, 2002.

When Mount Usu erupted on Hokkaido in 2000, the Japanese Ministry of Construction enlisted these UAVs to fly reconnaissance missions over the volcano's slopes to observe the advance of volcanic ash and mud threatening local villages. Manned aircraft were not allowed to fly these missions because of the danger posed by the volcano unexpectedly belching boulders and ash into the air. After a decade of consolidating a niche in Japan's agricultural industry, these robotic helicopters are poised to expand into civil roles, as the Mount Usu example demonstrates.

By contrast, there are some 160 U.S. UAVs either in use or undergoing evaluation, the overwhelming majority being owned by the military. Total U.S. military expenditures on UAVs surpassed $1 billion for the first time in 2003, making the U.S. the world leader in amount of money spent on UAVs. After the military, the next largest supporter of UAVs in the U.S. has been the National Aeronautics and Space Administration, which invested some $20 million annually in them under its Earth Research and Science Technology Program until 2003.

Western Europe collectively has the world's second largest fleet of UAVs, all of them dedicated to military reconnaissance. Three times the size of the U.S. UAV population, the European fleet is mainly divided among four countries, France, Germany, Great Britain, and Italy. Smaller numbers (one to two systems each) are operated by military units in Belgium, the Czech Republic, Denmark, Finland, the Netherlands, Sweden, Switzerland, and Greece. Finland has flown its UAVs in the dirty role of collecting air samples in the years following the Chernobyl reactor radiation spill.

INDUSTRY

Worldwide, 52 countries have an association with UAVs, whether as developers, manufacturers, operators, and/or exporters of them (see figures on page 133). Worldwide, there are some 250 models of UAVs to choose from, some with production histories approaching two decades to those that are only millions of dollars and several years from reality. Of these countries, 41 actively fly 80 of these 250 models of UAVs.

The United States, offering some 160 models, accounts for nearly two-thirds of those 250 UAV designs on the market, although many are admittedly far from production ready. At last count, there were 40 U.S. companies, ranging in size from giant Boeing to any of a dozen one- to ten-man shops existing on Small Business Innovative Research (SBIR) grants engaged in UAV development. As recently as the late 1990s, the distribution of these companies shown in the following figure was evenly strewn along the axis from the lower right to the upper left, but corporate acquisitions and mergers since then have thinned out the middle, lending more of a dumbbell shape to the distribution and leaving only a few prospects for future acquisition in the

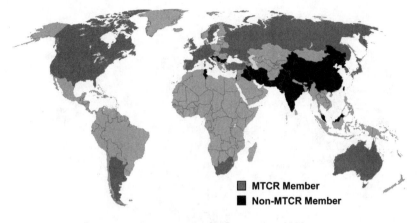

UAV manufacturing countries, 2002.

now-deserted middle. There is little correlation between size of the company and size of the UAV produced. Lockheed Martin, one of the largest, produces the 5-lb Sentry Owl, while Aurora, one of the smallest, produced the 140-ft wingspan Theseus. Their geographic distribution is very distinctly polarized between east and west coast, with a healthy cluster in Texas.

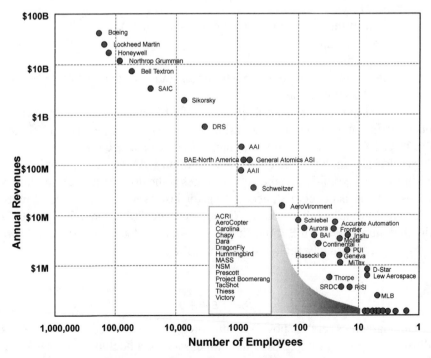

U.S. UAV industry revenues and employees.

Geographic distribution of the U.S. UAV industry, 2002.

MARKETS

UAV markets can be divided into five categories:
1) military
2) non-military government, or civil (such as NASA)
3) commercial
4) academic
5) nonprofit or nongovernmental organizations, such as the Red Cross or the United Nations

Of the 2400 UAVs described in the first figure as being in current operation around the world, 66% fall in the commercial category, 31% in the military, 2% in the civil, and less than 1% in academic circles.

Worldwide, in 2002, the annual spending on UAVs totals approximately $2 billion. Geographically, the United States is the dominant UAV market as far as funding is concerned, accounting for just over half of that total, with Western Europe another quarter. The remaining 20% is relatively evenly divided among Eastern Europe (including Russia), Israel, Japan, and Asia/Africa (predominately India and South Africa).

INFLUENCES

Demographics are a positive influence on unmanned aircraft, as is illustrated by UAVs being used to compensate for a declining labor pool on Japanese farms in the opening paragraphs of this chapter. Because UAVs fly,

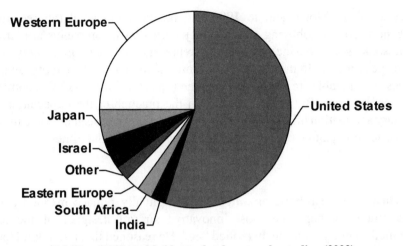

$2 billion UAV worldwide market by annual spending (2002).

some analysts think in terms of their major competition being pilots for future aviation jobs. The rationale is the large pool of Vietnam-era pilots is now retiring, a decline exacerbated by the post-cold war military draw down, so pilots will be in short supply, creating a demand for robotic aircraft. But this logic is flawed because the Vietnam bubble is unique to America, not worldwide, and the airlines have had a surfeit of pilots to hire from lately. It is better to think of UAVs as robots, and robots compete best for those jobs that are repetitive, hard, with a high pay-to-skill ratio, and perhaps dangerous. Admittedly, all flying sometimes possesses some of these attributes.

Economics is a widely anticipated positive influence, with the expectation that UAVs will be cheaper to build and operate than some manned aircraft, influencing the military to invest in their development. UAVs will not cost less to build than their manned equivalents, they will cost as much or even more. Ejection seats, oxygen regulators, and glass canopies are cheap items compared to ground stations and enough lines of software code and computing power to substitute for a pilot. Operating costs, however, may be another matter. Without the need to actually fly a plane to maintain pilot proficiency, the price to operate a fleet of UAVs may devolve to the electricity bill for running its combination ground control station/simulator. While training flights may be reduced by UAVs, they will never be eliminated, if for no other reason than to train human pilots to fly with them cooperatively.

Finally, politics is a fickle influence. At the national and international level, political considerations favorably influence UAVs. UAVs preclude the diplomatic difficulties posed by prisoners of war or scenes like those televised from Mogadishu in 1993. At the industrial level (appreciating that the legal and lobbying staffs of major aerospace companies are larger than some small businesses), UAVs represent both an opportunity and a business threat. If they can receive orders for them in quantity and at prices comparable to those for manned aircraft, then UAV contracts are opportunities. But that has not been the practice or the expectation of military acquisition officials, so UAVs have instead been seen by some major aerospace companies as threats to their established product lines.

SUMMARY

When asked if robotic aircraft would eventually replace manned ones, Burt Rutan, perhaps the most innovative aircraft designer of the late twentieth century, vigorously replied "no." He reasoned that if mankind took the aviator out of aviation, the romance of flying, the basic motivation that has driven aviation to improve itself over the past century, would be lost, and aviation, manned and unmanned, would dwindle and die. UAVs may replace manned aircraft in some roles where they possess superior capabilities, but overall, manned flight will endure and be complemented by unmanned flight.

Chapter 16

FUTURE HISTORY

TRENDS

The military market for UAVs has shown a strong positive trend since the end of the Cold War (1990) and this is expected to accelerate in the first decade of the twenty-first century. This is largely due to the outbreak of, and U.S. involvement in, numerous contingencies since then. Concurrently the size of the U.S. military has been drawn down over the same period, fanning a desire for increased reliance on robotics. The commercial market trend for robotics is also steadily upward, although not so strong. Technology has supported these trends, with the availability of ever cheaper and more capable microprocessors enabling these robotics. The major hurdles to the continuation of this trend are the increasing amount (and cost) of the software involved and the silicon chip manufacturing barrier, due to be reached in the 2015–2020 timeframe.

NEW TECHNOLOGIES

Throughout history, new technologies have been the enablers of new economies, and a number of new technologies are emerging at the dawn of the twenty-first century with great promise for aviation, unmanned and manned. The biological sciences have seen tremendous advances over the past decade, to the extent bioengineers and aerospace engineers may soon be working on common aircraft projects. The need for lighter, stronger aerostructures has led from wood to aluminum to titanium to composites. The next step may well be transgenetic biopolymers. A commercial company has succeeded in breeding goats with the silk-producing gene of spiders in their mammary glands and is productionizing the extraction of this silk from their milk. Spider silk has twice the tensile strength of steel yet is 25% lighter than carbon composites, and it is flexible. Imagine an aircraft skin made of it in which the servo actuators, hydraulics, electric motors, and control rods of today's aircraft control surfaces are replaced by the ability to warp wings and stabilizers by flexing their skin, much as the Wright brothers first conceived.

British bombers were suffering heavy losses in the early days of World War II, with many planes returning with severely damaged fuselages.

Losses continued at the same rate despite armor plating being added to the fuselages until someone pointed out it made more sense to place the armor where there was little or no damage in the returning planes. That was where the planes being lost were logically being hit. Composites have enabled lighter airframes, but the repair of damaged composites is far weaker than the original due to the loss of the material's originally plied construction, called aeroelastic tailoring. Researchers have recently devised a way to manufacture composite material with embedded microcapsules of "glue," so that any damage will open these capsules and heal the crack before it can propagate. This is known as an autonomic, or self-repairing, material, and such a material would be of great value in long endurance UAVs.

The antennas necessary for UAVs to communicate with their handlers may soon evolve from today's dish or blade to being conformal, then made of film, then sprayed on. Imagine an entire aircraft fuselage and/or wing that functions as an antenna, providing higher gain while eliminating the weight and power draw of present antenna drives.

Future UAVs will evolve from being robots operated at a distance to independent robots, able to self-actualize to perform a given task. This ability, autonomy, has many levels emerging by which it is defined, but ultimate autonomy will require capabilities analogous to those of the human brain by future UAV mission management computers. To achieve that level, machine processing will have to match that of the human brain in speed, memory, and quality of algorithms, or thinking patterns. Moore's Law predicts the speed of microprocessors will reach parity with the human brain around 2015. Others estimate the memory capacity of a PC will equal that of the human memory closer to 2030. As to when or how many lines of software code equate to "thinking" is still an open question, but it is noteworthy that pattern recognition by software today is generally inferior to that of a human.

As for the controller of UAVs, he will eventually be linked to his remote charge through his own neuromuscular system. Today's ground station vans are already being superceded by wearable harnesses with joysticks and face visors allowing the wearer to "see" through the UAV sensor, regardless of where he faces. Vests will soon provide him the tactile sensations "felt" by the UAV when it turns or dives or encounters turbulence. Eventually, UAV pilots will be wired so that the electrical signals they send to their muscles will translate into instantaneous control inputs to the UAV. To paraphrase a popular saying, the future UAV pilot will transition from seeing the plane to being the plane.

NEW ROLES

"New" roles for UAVs can be categorized as either moving into those roles currently performed by manned aircraft or those not performed by any aircraft. In either case, the UAV must offer some unique incentive, financial or capability-wise, in order to gain acceptance in that role. An example of the former is robotic air transport of cargo, and eventually people, and of the latter, that topical role, countering terrorism.

The express delivery airlines led the way in downsizing from the three-man cockpit to two during the 1980s, paying the up-front cost to convert former passenger airliners to this leaner mode of operation. Once the concept was proven and new regulations established, new production passenger airliners followed suit. The motivation then was to reduce operating costs by reducing labor. The future motivation will become reducing operating losses by eliminating aircrew strikes.

Commercial robotic flight, for both cargo and passengers, is essentially already here. Within 20 min after takeoff, today's pilots put airliners on autopilot and let it fly the plane along its preprogrammed route, turns and all, during the cruise portion of the flight. If there is bad weather at the destination, the pilot may elect to have it fly an autopilot-coupled approach right down to touchdown. This is occurring routinely, every day, worldwide because it is safer.

How will commercial robotic flight occur? Gradually. At some point, the cargo carrier's two-man cockpit will become a one-man cockpit and then a no-man one. Over the course of a decade, the media and the flying public will notice that cargo planes are not crashing as often or as spectacularly as ones carrying passengers. Unions allowing, airlines will then transition from two- to one-man cockpits. Eventually, that one pilot will be there just to pre-program the autopilot and smile at the boarding passengers before stepping off discretely just before push-back. Unless you are sitting in the first row of first class, you will not even notice.

For fighting terrorism, unmanned aviation (and robotics in general) offer three unique virtues over their manned counterparts. First, the terrorist scores a propaganda "win" if he manages to kill one person or down one aircraft or ship, regardless of his own losses (USS *Cole*, 2000). However, the act of destroying robotic aircraft presents the terrorist with more of a loss than a win because it reveals his location, consumes his relatively meager supply of high-end weaponry, and alerts his real prey, the manned aircraft or ship, to be more proactive in the future—all without gaining a propaganda win.

Second, he gains a political bargaining chip if he is able to take people hostage or capture downed pilots and either exchange them for captured terrorists (Kandahar, 1999) or kill them to turn public opinion (and political will) against involvement in their region (Somalia, 1993). With no one onboard, the debris of crashed UAVs offer him nothing with which to bargain. A new Predator® can be built in six weeks; a new combat pilot requires two years to be fully trained.

Third, targeting terrorists involves long, dull periods of uninterrupted surveillance of numerous sites, tying up forces assigned to such duty, placing them in dangerous locations, and, as inactivity drags on, sapping their morale. Robotic surveillance can be continuous, relentless, and constantly alert. Just as knowledge of reconnaissance satellite pass times by terrorists cause them to restrict their activities for a few times each day for a few moments, surveillance by long endurance UAVs can reverse this schedule, restricting their activities to a few moments each day—unpredictable moments at that. Further, the current sensors carried by UAVs offer the added advantages of better resolution (by being closer), video, color, and real-time surveillance, all of which aid in recognizing and reacting to terrorists more quickly and with higher confidence. Such is the salient attribute of the Predator® UAV/ Hellfire missile combination used in Afghanistan.

CHALLENGES

To develop further, unmanned aviation, and specifically unmanned aerial vehicles (UAVs), needs improved reliability, a regulatory infrastructure, and a stable customer base, in that order.

The reliability of UAVs is vitally important because it underlies their affordability (an acquisition issue), their mission availability (an operations and logistics issue), and their acceptance into civil airspace (a regulatory issue). The mishap rates of currently fielded military UAVs are one to two orders of magnitude poorer than those of manned military aircraft and another order of magnitude poorer than that of airliners. Improved reliability is key to enhancing UAV affordability, availability, and acceptance.

AFFORDABILITY

More reliable UAVs are more expensive UAVs, so enhancing reliability must be weighed as a trade-off between increased up-front costs for a given UAV and reduced operating and maintenance costs over the system's lifetime. The reliability of UAVs is closely tied to their affordability primarily because the public has come to expect them to be less expensive than their manned counterparts. This expectation is based on the UAV's generally smaller size and the omission of those systems needed to support a pilot or aircrew. However, beyond these two measures, other cost-saving measures to enhance affordability begin to impact reliability. Improved reliability can offer lower operating costs by reducing maintenance man-hours per flight hour (MMH/FH) and by decreasing the number of spares and attrition aircraft procured.

AVAILABILITY

With the removal of the pilot, the rationale for including the same level of redundancy, or for using man-rated components considered crucial for his safety, can go undefended in UAV design reviews and may be sacrificed for affordability. Less redundancy and lower-quality components, while making UAVs even cheaper to produce, mean they become more prone to in-flight loss and more dependent on maintenance, both impacting their availability and ultimately their life-cycle cost (LCC).

ACCEPTANCE

Finally, improving reliability is key to winning the confidence of the general public, the acceptance of other aviation constituencies (airlines, general aviation, business aviation, and so forth), and the willingness of governmental authorities to regulate UAV flight.

Regulation of UAVs is important because it will provide a legal basis for UAV operation in the National Aerospace System for the first time. A legal basis is necessary for determining liability in accidents and, therefore, insurability for UAV operators. Regulation of UAV operations also should lead to acceptance by international civil aviation authorities (ICAO) of UAV operations. Such acceptance will greatly facilitate obtaining overflight and landing privileges when foreign flights are required. In addition, regulation will save time and resources by providing one standardized, rapid process for granting flight clearances. Third, regulation will encourage the use of UAVs in civil and commercial applications, resulting in potentially lower production costs for both the civilian and military markets.

Finally, the UAV industry needs to convince an established commercial market to adopt UAVs as a better alternative to their current way of doing business (for instance, using UAVs as pseudosatellites in lieu of cell towers). Congressional recognition in the form of an unmanned aviation act would help. Good precedents exist. The Air Mail Act of 1925 transferred what had been a federal function since 1918 into the commercial sector. The first commercial airmail contract flight followed in 1926 and by 1927 all airmail was carried under contract. These contracts were crucial to funding the start of what became the U.S. airline industry. The Commercial Space Act of 1997 sought to accomplish much the same, to transition a heretofore government function into the private sector by changing the federal government's role from being a competitor to a fledgling industry to becoming a customer for its services. When the words "unmanned aviation" find their way into similar Congressional language, UAVs will have a start at becoming full-fledged players in aviation markets.

SUMMARY

The development of unmanned aviation has been the driving or contributing motivation behind many of the key innovations in aviation: the autopilot, the inertial navigation system, and data links, to name a few. Although it was hobbled by technology insufficiencies through most of the twentieth century, focused efforts in small, discrete military projects overcame the problems of automatic stabilization, remote control, and autonomous navigation. The last several decades have been spent improving the technologies supporting these capabilities largely through the integration of increasingly capable microprocessors in the flight and mission management computers flown on UAVs. The early part of the twenty-first century will see even more enhancements in UAVs as they continue their growth. The ongoing revolution in the biological sciences will eventually impact aviation and, together with future microprocessors, will enable intelligent, vice robotic, UAVs to fly over the Earth and other planets.

SIGNIFICANT DATES IN UNMANNED AVIATION

1804 First flight by a fixed wing unmanned model glider; performed by George Cayley in Yorkshire, England.

1848 First powered flight by an unmanned aircraft; performed by John Stringfellow with his 12-foot wingspan model Aerial Steam Carriage at Chard, England.

2 May 1857 Flight by a powered, unmanned aircraft, by Du Temple in France.

May 1898 First public demonstration of radio-control applied to a moving vehicle; performed by Nikola Tesla with his Telautomaton RC boat in New York's Madison Square Garden.

7 January 1918 First production contract awarded for an unmanned aircraft; awarded by the U.S. Army to the Dayton Wright Airplane Company for 25 Liberty Eagles.

6 March 1918 First successful flight of a powered, full-size, unmanned aircraft; performed by the Curtiss Sperry Aerial Torpedo (based on the Curtiss Speed Scout design), which flew 1000 yd before landing and being recovered off Copiague, Long Island, New York.

4 October 1918 First successful flight by a production unmanned aircraft; performed by the Dayton Wright Liberty Eagle from Dayton's South Field to near Xenia, Ohio.

1922 First launch of an unmanned aircraft (RAE 1921 Target) from an aircraft carrier (HMS Argus).

3 September 1924 First successful flight by radio controlled unmanned aircraft without a safety pilot onboard; performed by the British RAE 1921 Target 1921, which flew 39 minutes.

1933   First use of an unmanned aircraft as a target drone; performed by a Fairey Queen for gunnery practice by the British Fleet in the Mediterranean Sea.

12 June 1944   First combat use of unmanned aircraft (German Fi-103 "V-1") in the cruise missile role.

19 October 1944   First combat use of unmanned aircraft (U.S. Navy TDR-1 attack drone) in the strike role, dropping ten bombs on Japanese gun positions on Ballale Island.

April 1946   First use of unmanned aircraft for science research; performed by a converted Northrop P-61 Black Widow for flights into thunderstorms by the U.S. Weather Bureau to collect meteorological data.

1955   First flight of an unmanned aircraft designed for reconnaissance; performed by the Northrop Radioplane SD-1 Falconer/Observer, later fielded by the U.S. and British armies.

12 August 1960   First free flight by an unmanned helicopter; performed by the Gyrodyne QH-50A at NATC Patuxent River, Maryland.

20–21 August 1998   First trans-Atlantic crossing by an unmanned air-craft; performed by the Insitu Group's Aerosonde™ *Laima* between Bell Island, Newfoundland, and Benbecula, Outer Hebrides, Scotland.

22–23 April 2001   First trans-Pacific crossing by an unmanned aircraft; per-formed by the Northrop Grumman Global Hawk "Southern Cross II" between Edwards AFB, California, and RAF Edinburgh, Australia.

Appendix B

NOTABLE PERSONALITIES IN UNMANNED AVIATION

Boucher, Robert J. Born 12 July 1932 in Willimantic, Connecticut. Graduated from the University of Connecticut (1954) and Yale University (1955) in electrical engineering, then worked for Hughes Aircraft Company until 1973 before founding Astro Flight, Inc. Developed and flew the world's first solar-powered aircraft, the *Sunrise I* in 1974.

Cayley, Sir George Born 27 December 1773 in Yorkshire, England. Developed the principles and concept behind the modern airplane and flew the first unmanned, unpowered aircraft in 1804. Died 15 December 1857 at Brompton, Yorkshire, England.

Cody, Samuel See Cowdery, Samuel Franklin.

Cowdery, Samuel Franklin Born 1867 in Davenport, Iowa. Credited with developing and flying the first indigenously designed British airplane and perhaps the first British unmanned aircraft, a powered kite, in 1908. Died 7 August 1913 in the UK.

Curtiss, Glen Hammond Born 21 May 1878 in Hammondsport, New York. Designed and built the Speed Scout, adapted as the first powered full size unmanned aircraft to fly, the Curtiss Sperry Aerial Torpedo in 1918. Died 23 July 1930.

Denny, Reginald Leigh Born 20 November 1891 in Richmond, Surrey, England. Introduced radio-controlled airplanes to the public as a hobby and to the U.S. Army as target drones, his design of which led to the first use of unmanned aircraft in the reconnaissance role (U.S. Army's SD-1 Observer, 1955). Died 16 June 1967 in Richmond, Surrey, England.

Draper, Charles Stark "Doc" Born 2 October 1901 in Windsor, Missouri. Developed the inertial navigation system (INS) that enabled accurate, sustained autonomous navigation by unmanned aircraft. Died 25 July 1987.

Fahrney, Delmer Stater   Born 23 October 1898 in Grove, Oklahoma, Indian Territory. Graduated from the U.S. Naval Academy in 1920 and retired as a rear admiral in 1950. Oversaw the development of radio-controlled aircraft and introduced them in the role of target drones for the U.S. Navy in 1939. Died 12 September 1984 in La Mesa, California.

Hewitt, Peter Cooper   Born 5 May 1861. Conveyed the concept of an aerial torpedo from Tesla to Elmer Sperry, then helped Sperry finance its development during World War I. Died 25 August 1921 in Brooklyn, New York.

Kettering, Charles Franklin "Boss"   Born 29 August 1876 in Loudenville, Ohio. Graduated from Ohio State University in 1904. Developed the first unmanned aircraft to enter production (U.S. Army Liberty Eagle, 1918). Died 25 November 1958.

Low, Archibald Montgomery   Born 17 October 1888 in Scotland, UK. Educated at Skerry's College in Glasgow. Invented first radio control equipment intended for use controlling an unmanned aircraft in 1916–17. Served as President of the British Institute of Engineering Technology and wrote science fiction stories in his later life. Died 18 September 1956.

Morgan, W. Ray   Born 25 January 1947 in Mount Airy, North Carolina. Graduated from North Carolina State University (1969) in aerospace engineering, then worked for Lockheed California Company until 1980 and AeroVironment until 2000. Led development of the series of solar-powered flying wings (HALSOL, *Pathfinder*, *Pathfinder Plus*, *Centurion*, and *Helios*) that proved the concept of unfueled, high-altitude endurance UAVs.

Norden, Carl Lukas   Born 23 April 1880 in Semerang, Java, Indonesia. Immigrated to the United States in 1904. Developed the flywheel catapult that enabled successful launches of early Sperry unmanned aircraft. Later helped the U.S. Navy develop radio controls for airplanes. Emigrated to Switzerland after World War II, where he died 1965.

Sperry, Elmer Ambrose   Born 12 October 1860 in Cortland, New York. Attended Cornell University for one year. Developed the gyrostablizer, critical for maintaining unmanned aircraft in straight and level flight without an onboard pilot. Helped develop the first two powered unmanned aircraft, the Curtiss Sperry Aerial Torpedo and the Dayton Wright Liberty Eagle. Died 16 June 1930.

Sperry, Lawrence Burst Born 21 December 1893 in Chicago, Illinois. Son of Elmer Sperry. Served as test pilot for the first powered unmanned aircraft, the U.S. Navy Curtiss Sperry Aerial Torpedo, and for tests of the U.S. Army's radio controlled Messenger Aerial Torpedo. Died 13 December 1923 when his airplane crashed into the English Channel.

Tesla, Nikola Born 9 July 1856 in Smiljan, Croatia. Immigrated to the United States in 1884. Developed and demonstrated the first radio controlled vehicle, a four-foot boat, at an electrical exposition in New York's Madison Square Garden in 1898. Died 7 January 1943 in New York.

Stringfellow, John Born 1799 in Chard, Somerset, England. Developed and flew the world's first powered unmanned aircraft, a steam-powered model with a 12-foot wingspan, in 1848. Died 1883.

Museums with Unmanned Aircraft

Unmanned Aircraft on Display	Museum References
Beechcraft AQM-37A	12
Beechcraft KDB-1	12
Boeing AGM-86B ALCM	9
Boeing Brave 300	12
Boeing Brave 200	12
Boeing Condor	2
Curtiss N-9	6
Curtiss Sperry Aerial Torpedo	1
Dayton Wright Liberty Eagle	9
DeHavilland Tiger Moth (Queen Bee)	3
Fairey IIIF (Queen)	3
General Dynamics BGM-109 Gryphon	9
General Dynamics AGM-129 ACM	9
Globe KD6G-2	12
Gyrodyne QH-50C	8, 13, 14, 15, 16, 17, 18, 19, 20
Insitu Aerosonde *Laima*	4
Lockheed D-21B	4, 8, 9
Northrop MQM-57	8
Northrop AGM-126 Tacit Rainbow	9
PUI RQ-2A Pioneer	5, 7, 10
Radioplane RP-5A	11
Sperry M-1 Messenger (MAT)	1, 9
Teledyne Ryan AQM-34L	8
Teledyne Ryan BQM-34 Firebee	12
Teledyne Ryan YQM-98A Tern	8

1) Cradle of Aviation Museum, Museum Lane, Mitchell Field, Garden City, NY 11530, 516-222-1190, www.cradleofaviation.org

2) Hiller Aviation Museum, 601 Skyway Road, San Carlos, CA 94070, 650-654-0200, www.hiller.org

3) Imperial War Museum, Duxford, Cambridgeshire, UK CB2 4QR, +44 1223 835 000, www.iwm.org.uk/duxford

4) Museum of Flight, 9404 East Marginal Way S., Seattle, WA 98108, 206-764-5720, www.museumofflight.org

5) National Air & Space Museum, 6th St and Independence Ave., Washington, DC 20560, 02-357-2700, www.nasm.si.edu

6) National Museum of Naval Aviation, 1750 Radford Blvd., Pensacola, FL 32508, 850-452-2311, www.naval-air.org

7) Nauticus, 1 Waterside Drive, Norfolk, VA 23510, 800-664-1080, www.nauticus.org

8) Pima Air & Space Museum, 6000 E. Valencia Road, Tucson, AZ 85706, 602-574-9658, www.pimaair.org

9) US Air Force Museum, 1100 Spaatz Street, Wright-Patterson AFB, OH 45433, 937-255-3286, www.wpafb.af.mil/museum

10) USS Missouri Memorial, POB 6339, Pearl Harbor, Hawaii 96818, 877-644-4896, www.ussmissouri.com

11) Western Museum of Flight, 12016 Prairie Ave., Hawthorne, CA 90250, 310-332-6228, www.wmof.com

12) Yanks Air Museum, Chino, CA, 909-597-1734, www.yanksair.com

13) American Helicopter Museum & Education Center, 1220 American Blvd., West Chester, PA 19380, 610-436-9600, www.helicopter museum.org

14) New England Air Museum, Bradley International Airport, Windsor Locks, CT, 860-623-3305, www.neam.org

15) USS Joseph P. Kennedy Jr. DD-850 Destroyer Museum, Battleship Cove, Fall River, MA, 508-678-1100, www.ussjpkennedyjr.org

16) Patriots Point Naval & Maritime Museum, 40 Patriots Point Rd., Mount Pleasant, SC 29464, 800-248-3508, www.state.sc.us/patpt

17) Carolinas Aviation Museum, 4108 Airport Dr., Charlotte, NC 28208, 704-359-8442

18) USS Radford National Naval Museum, 228 W. Canal St., Newcomerstown, OH, 740-498-4446, www.ussradford446.org

19) USS Orleck DD-886, Orange, TX 77631, 409-883-8346, www.ussorleck.org

20) Gyrodyne Helicopter Historical Foundation Aircraft Museum, Reno-Stead Airport, Reno, NV 89506, 775-329-1214, www.gyrodynehelicopters.com

NOTABLE UNMANNED AIRCRAFT

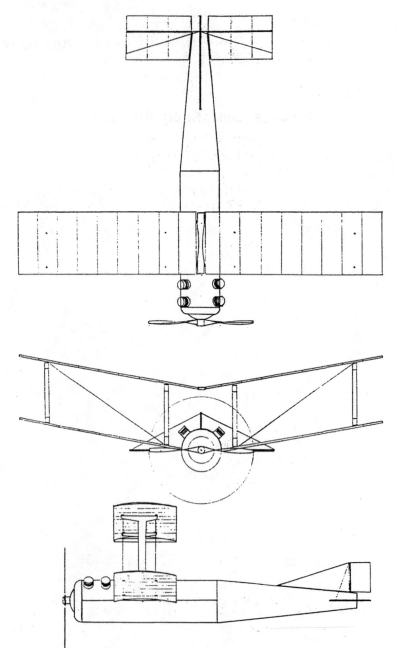

First Production Unmanned Aircraft
U.S. Army Liberty Eagle (Kettering *Bug*)
Dayton Wright Airplane Co.

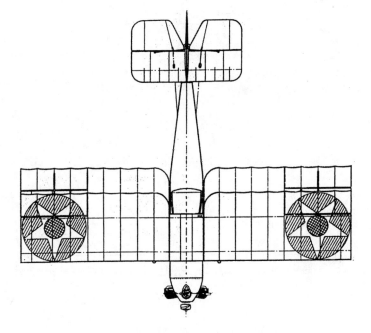

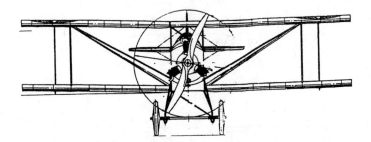

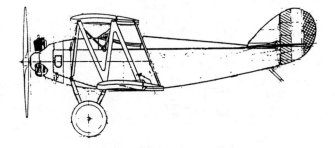

**First Radio-Controlled Demonstration Aircraft
U.S. Army Messenger Aerial Torpedo (MAT)
Sperry Airplane Co.**

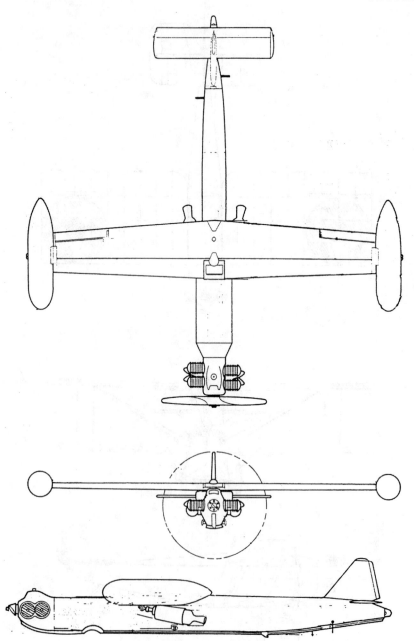

First Reconnaissance UAV
U.S. Army SD-1 Observer
Northrop Radioplane Div.

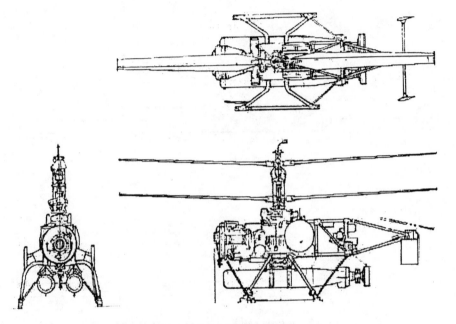

First Unmanned Rotorcraft
U.S. Navy QH-50C DASH
Gyrodyne Corp.

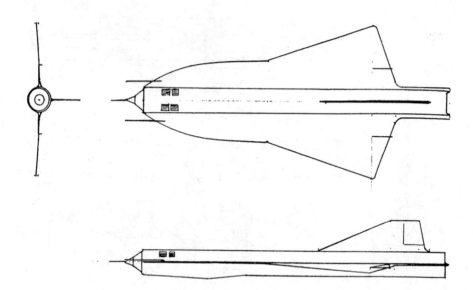

Holder of the UAV Speed Record (until NASA X-43 Hyper-X)
U.S. Air Force GTD-21B
Lockheed Skunk Works Co.

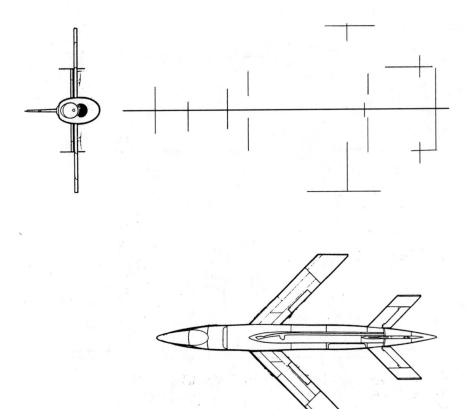

Workhorse of the Vietnam War
U.S. Air Force AQM-34L
Teledyne Ryan

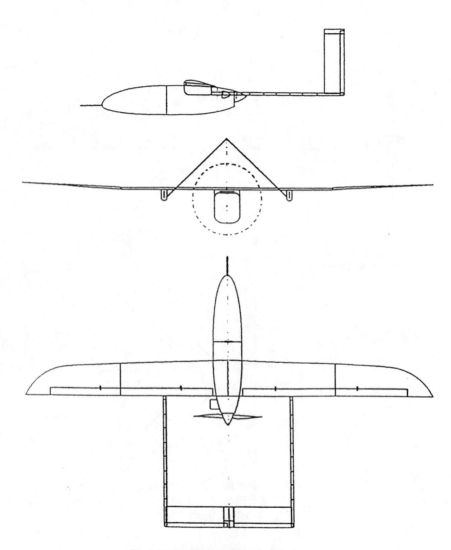

First UAV to cross the Atlantic Ocean
Aerosonde *Laima*
The Insitu Group

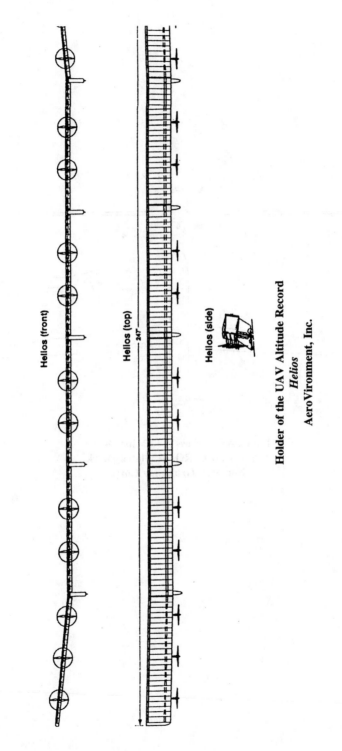

Helios (front)

Helios (top)

247

Helios (side)

Holder of the UAV Altitude Record
Helios
AeroVironment, Inc.

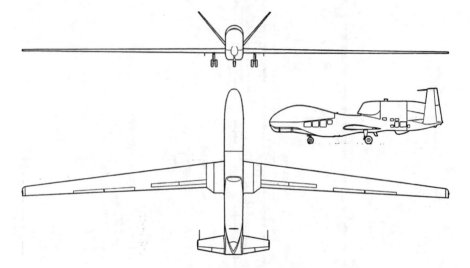

First UAV to cross the Pacific Ocean
U.S. Air Force RQ-4A Global Hawk
Northrop Grumman Corp.

UAV CHARACTERISTICS AND PERFORMANCE TABLE

UAV Engine(s)	No. Built	First Flight	Span	Length	Weights (lb)			Speed (kt)	Ceiling (ft)	Endurance (h)
					Gross	Payload	Empty			
Dayton Wright Liberty Eagle 1 × 37 hp DePalma Recip	36	1918	15' 0"	12' 6"	530	180	—	100	12,000	1
Sperry M-1/MAT Messenger 1 × 60 hp Lawrence L-4 Radial	42	1920	20' 0"	17' 9"	862	150	623	80	—	<2
Northrop SD-1 Falconer Observer 1 × 72 hp McCulloch 0-100-2 Recip	1445	1955	13' 5"	13' 5"	466	75	354	160	15,000	0.6
Gyrodyne QH-50C DASH 1 × 300 hp Boeing T50 Turboshaft	756	1960	20' 0"	12' 11"	2281	866	1100	80	16,200	3.9
Lockheed GTD-21B 1 × 12,000 lbst Marquardt RJ43 Ramjet	38	1966	19' 0"	43' 2"	20,000	—	11,200	M3.35	90,000+	2
AeroVironment Helios 14 × 2 hp Electric Motors	1	1999	247' 0"	12' 0"	2,048	726	1,322	22–140	100,000	17+
Insitu Group Aerosonde 1 × 1 hp Enya R120 Recip	50	1995	9' 6"	6' 2"	29.5	1	18	56	14,800	32
Northrop Grumman RQ-4A Global Hawk 1 × 7,600 lbst Allison AE3007H Turbofan	6*	1998	116' 2"	44' 5"	25,600	1,960	8,530	345	66,000	32
Teledyne Ryan AQM-34L 1 × 1,920 lbst Teledyne CAE J69 Turbojet	250+	1968	13'	29"	3,065	965	1,500	560	60,000	1.25

*Number built as of the end of 2002. Currently in production with 51 planned for the U.S. Air Force.

BIBLIOGRAPHY

Sources marked with an asterisk (*) were provided by Mr. Brett Stolle from the archives of the Research Division of the U.S. Air Force Museum, Dayton, Ohio, and whose contribution is gratefully acknowledged.

Anderson, J. D., Jr., *Introduction to Flight*, McGraw-Hill, New York, 1989.

Armitage, M., *Unmanned Aircraft*, Brassey's Defence Publishers, London, 1988.

Botzum, R. A., *50 Years of Target Drone Aircraft*, Northrop, Newbury Park, CA, 1985.

Brunn, B. E., Interview, Norfolk, VA, March, 2000.

Charon, L. P., "Front-Line Photo Drone Ready for Robot Recon," *Marine Corps Gazette*, Vol. 50, No. 8, August 1966, p. 1612.

Copp, D. S., *A Few Great Captains*, EPM Publications, McLean, VA, 1980.

Crabb, J., Interview, San Antonio, TX, June 2000.

Dale, J., Interview, Portsmouth, VA, May 2002.

*Dayton Metal Products Company, Research Division, "Description and Partial History of the 'Kettering' Torpedo-Airplane," undated.

D'Alto, N., "Victorian Dreams of Flight," *Aviation History*, January 2004, pp. 42–48.

Fiszer, M. and Gruszczynski, J., "Russian UAV Programs at Turning Point," *The Journal Of Electronic Defense*, Vol. 26, No. 4, April 2003, p. 36.

Freedman, R., *The Wright Brothers: How They Invented the Airplane*, Holiday House, New York, 1991.

*Gearhart, G., II, "Resume of Aerial Torpedo Development," 10 April 1926.

Gorn, M. H., *Prophecy Fulfilled: "Towards New Horizons" and Its Legacy*, Air Force History and Museums Program, 1994.

Hallstead, W. F., "The U.S. Navy's Kamikazes," *Aviation History*, January 2004, pp. 50–58, 79.

Howeth, L. S., *History of Communications-Electronics in the United States Navy*, Bureau of Ships and Office of Naval History, Washington, DC, 1963.

Josephy, A. M., Jr., *The American Heritage History of Flight*, Simon & Schuster, New York, 1962.

Lynn, L. and Entzminger, J., "Are UAVs Finally Going to Make It—and Why? (Recent History of Endurance UAVs)," Presentation to DARPA-Tech 2002 Symposium, San Diego, July 2002.

Munson, K., *World Unmanned Aircraft*, Jane's Publishing, London, 1988.

Pearson, L., "Developing the Flying Bomb," *Naval Aviation in World War I*, edited by Adrien D. van Wyen, U.S. Government Printing Office, Washington, DC, 1969, pp. 70–74.

Peebles, C., *Dark Eagles: A History of Top Secret U.S. Aircraft Programs*, Presidio Press, Novato, CA, 1995.

Phillips, W., "Mr. Sperry's Mighty Midget," *Air Classics*, Vol. 15, No. 5, May 1979, pp. 14–23.

Schoening, J. P., series of interviews, Lakewood, NJ, Aug.–Sept., 2002.

Seifer, M. J., *Wizard: The Life and Times of Nikola Tesla*, Birch Lane Press, Secaucus, NJ, 1996.

Tart, L. and Keefe, R., *The Price of Vigilance*, Ballantine, New York, 2001.

Taylor, J. W. R. (ed), *Jane's All The World's Aircraft, 1968–69*, BPC Publishing, London, 1968.

Thirtle, M. R., Johnson, R. V., and Birkler, J. L., *The Predator ACTD: A Case Study for Transition Planning to the Formal Acquisition Process*, RAND, Santa Monica, CA, 1997.

Tsach, S. and Dubester, Y., "25 Years of UAV Development in IAI: Lessons and Future Directions," unpublished paper, 2002.

Van Riper, P. K., "Unmanned Aerial Vehicle Programs, U.S. Marine Corps," Congressional Hearings, Intelligence and Security, 1997.

*Wiggin, C. and Eisenberg, H., "Top Secret Revealed! Our 1918 Missile," *Saga*, August, 1961, pp. 17–21, 93–95.

www.gyrodynehelicopters.com. Gyrodyne Helicopter Historical Foundation, Reno, NV, May 2003.

INDEX

SUPPORTING MATERIALS

A complete listing of AIAA publications is available at http://www.aiaa.org